Investigate the Possibilities

Elementary Earth Science

THE EARTH
Its Structure & Its Changes

Teacher's Guide

**Tom DeRosa
Carolyn Reeves**

THE EARTH
Its Structure & Its Changes

Tom DeRosa
Carolyn Reeves

Teacher's Guide and Student Journal

First printing: January 2011

Master Books® is a division of the New Leaf Publishing Group, Inc.

ISBN: 978-0-89051-592-1

Cover by Diana Bogardus

Unless otherwise noted, Scripture quotations are from the New International Version of the Bible.

Please consider requesting that a copy of this volume be purchased by your local library system.

Printed in the United States of America

Please visit our website for other great titles:
www.masterbooks.net

For information regarding author interviews, please contact the publicity department at (870) 438-5288

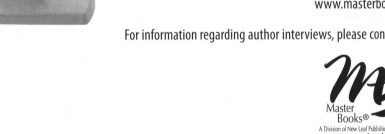

Master Books®
A Division of New Leaf Publishing Group
www.masterbooks.net

Teacher's Guide Table of Contents

Student Journal Table of Contents

The overall goal for each workbook is to include three components: good science, creation apologetics, and Bible references. This goal underlines the rationale for the design of the workbooks.

Science is a great area to teach, because children have a natural curiosity about the world. They want to know why and how things work, what things are made of, and where they came from. The trick is to tap into their curiosity so they want to find answers.

Many elementary-level science lessons begin with definitions and scientific explanations, followed by an activity. A more effective method is to reverse this order and begin with an activity whenever possible. The lessons found in these workbooks begin with an investigation, followed by scientific explanations and opportunities to apply the knowledge to other situations.

In addition to the investigations, there are sections on creation apologetics, written mostly in narrative forms; connections to Bible references; on-your-own challenges; pause and think questions; projects and contests; and historical stories about scientists and engineers. These sections encourage students to think more critically, to put scientific ideas into perspective, to learn more about how science works, to gain some expertise in a few areas, and to become more grounded in their faith in the Bible.

It is not expected that students will do everything suggested in the workbooks. The variety provides students with choices, both in selection of topics and in learning styles. Some students prefer hands-on activities and building things, while others prefer such things as writing, speaking, drama, or artistic expressions. Once some foundational ideas are in place, having choices is a highly motivating incentive for further learning.

Every effort has been made to provide a resource for good science that is accurate and engaging to young people. Most of the investigations were selected from lessons that have been tested and used in our Discovery classrooms. Careful consideration was given to the National Science Education Standards.

Format for Individual Lessons:

1. **Think about This:** The purpose of this section is to introduce something that will spark an interest in the upcoming investigation. Lesson beginnings are a good time to let students make observations on their own; for a demonstration by the teacher; or to include any other kind of engaging introduction that causes the students to want to get answers. Teachers should wait until after students have had an opportunity to do the investigation before answering too many questions. Ideally, lesson beginnings should stimulate the students' curiosity and make them want to know more. Lesson beginnings are also a good time for students to recall what they already know about the lesson topic. Making a connection to prior knowledge makes learning new ideas easier.

2. **The Investigative Problem:** This section brings a focus to the activity students are about to investigate and states the objectives of the lesson. Students should be encouraged throughout the investigation to ask questions about the things they want to know. It is the students' questions that connect with the students' natural curiosity and makes them want to learn more. Teachers should stress to students at the start of each lesson that the goal is to find possible solutions for the investigative problem.

3. **Gather These Materials:** All the supplies and materials that are needed for the investigation are listed. The teacher's book may contain additional information about substituting more inexpensive or easier to find materials.

4. **Procedures and Observations:** Instructions are given about how to do the investigation. The teacher's book may contain more specifics about the investigations. Students will write their observations as they perform the activity.

5. **The Science Stuff:** It is much easier for students to add new ideas to a topic in which they already have some knowledge or experience than it is to start from scratch on a topic they know nothing about. This section builds on the experience of the investigation.

6. **Making Connections:** Lessons learned become more permanent when they are related to other situations and ideas in the world. This section reminds students of concepts and ideas they likely already know. The scientific explanation for what the students observed should be more meaningful if it can be connected to other experiences and/or prior knowledge. The more connections that are made, the greater the students' level of understanding will become.

7. **Dig Deeper:** This section provides ideas for additional things to do or look up at home. Students will often want to learn more than what was in the lesson. This will give them some choices for further study. Students who show an interest in their own unanswered questions should be allowed to pursue their interests, provided the teacher approves of an alternative project. Students should aim to do at least one project per week from Dig Deeper or other project choices. The minimum requirements from this section should correspond to each student's grade level. Students may want to do more than one project from a lesson and none from other lessons. Remember, this is an opportunity for students to choose topics that they find interesting.

8. **What Did You Learn?** This section contains a brief assessment of the content of the lesson in the form of mostly short-answer questions.

9. **The Stumpers Corner:** The students may write two things they would like to learn more about or they may write two "stumper" questions (with answers) pertaining to the lesson. Stumper questions are short-answer questions to ask to family or classmates, but they should be hard enough to be a challenge.

NOTE TO THE TEACHER

The books in this series are designed to be applicable mainly for grades 3–8. The National Science Education Standards for levels 5–8 were the basis for much of the content. Recommendations for K–4 were also considered, because basic content builds from one level to another.

We feel it is best to leave grading up to the discretion of the teacher. However, for those who are not sure what would be a fair way to assess student work, the following is a suggestion.

1. Completion of 20 activities with write-up of observations — 30%

2. Completion of What Did You Learn Questions + paper and pencil quizzes — 35%

3. Projects, Contests, and Dig Deeper — 35%

The teacher must set the standards for the amount of work to be completed. The basic lessons will provide a solid foundation for each unit, but additional research and activities are a part of the learning strategy. The number of required projects should depend on the age, maturity, and grade level of the students. All students should choose and complete at least one project each week or 20 per semester. Fifth and sixth graders should complete 25 projects per semester. A minimum guide for seventh and eighth graders would be 30 projects. The projects can be chosen from "Dig Deeper" ideas or from any of the other projects and features. Additional projects would give extra credits. By all means, allow students to pursue their own interests and design their own research projects, as long as you approve first. Encourage older students to do the more difficult projects.

Students should keep their work in the Student Journal. If additional space is needed, teachers can provide files or notebooks to organize their work. You may or may not wish to assign a grade for total points, but a fair evaluation would be three levels, such as: minimum points, more than required, and super work. Remember, the teacher sets the standards for evaluating the work.

Ideally, if students miss a lab, they should find time to make it up or do one of the alternative activities. When this is not practical, make sure they understand the questions at the end of the lesson and have them do one of the "Dig Deeper" projects or another project.

You should be able to complete most of the 20 activities in a semester. Suppose you are on an 18-week time frame with science labs held once a week for two or three hours. Most investigations can be completed in an hour or less. Some of the shorter activities can be done on the same day or you may choose to do a teacher demonstration of a couple of the investigations.

It is suggested that at least five hours a week be allotted to the investigations, contests, sharing of student projects, discussion of "What Have You Learned" questions, and research time. More time may be needed for some of the research and projects.

Most of the equipment for these investigations can be obtained from hardware stores, grocery stores, and other local stores. You may want to look over the needed materials list now and begin to collect these items. For example, there is no need to make more than one trip to the hardware store if you know what you will need for this unit.

Note that the scientists introduced in this book, along with their life span, are listed in the Introduction. A useful tool is to make is a time-line. Scientists from other books in this series should be part of this time-line. A long chart about 30 centimeters (12 inches) high and about 5 meters (16 feet) long works well. The chart should be divided into equal 100-year intervals beginning with 2000 B.C. This chart can be displayed as needed. As each new scientist is introduced, a colored block can be taped to the chart showing the life span of the scientist/technician with the name of the person written inside. It is extremely helpful to students to relate other historical people to this time-line. Science is not an isolated endeavor that occurs apart from other historical events and should be shown in the context of what is happening in the culture and the world.

Orange You Going to Map the Earth?

Think about This Dustin and Elizabeth were trying to find the countries of Togo and Greenland on the globe. "Look, Togo is in Africa next to the Atlantic Ocean, a little north of the equator, so I guess it's pretty hot there," Dustin said as he finally located the little country.

"Greenland is easy to find, and it is probably much colder than Togo. I wonder why it looks bigger on the wall map than it does on the globe?" Elizabeth said.

Have you ever wondered how a round globe of the earth could show the same countries and oceans as a flat map of the earth? Have you ever wondered how much information you could find out about different countries from looking at a map? Let's find out some answers!

The Investigative Problems
How can the countries on a round earth be shown on a flat map?
What do lines on the map tell us?

Procedure & Observations

Part A

1. Draw a line from the flower end to the stem end of the orange with the black marker to represent a longitudinal line going from the North Pole to the South Pole. Draw five more longitudinal lines in the same way, keeping them spaced about the same distance apart.

2. Now make several dots with your marker halfway between the "poles" and connect the dots by drawing a line around the center of the orange to represent the equator. Draw a few more lines above and below this line to represent latitudinal lines. Make them parallel to the "equator." Notice that these lines get smaller and smaller as you move toward the "poles."

3. Draw some landmasses with your green marker to represent continents. Don't worry about drawing continents with accurate sizes and shapes.

6

Gather These Things:
✓ Whole orange
✓ Fruit peeler
✓ Plastic knife
✓ Black marker
✓ Green marker
✓ Flat map of the world
✓ Globe (if available)

4. Cut the skin of the orange with the fruit peeler along each "longitudinal line" down to within one centimeter (just under half an inch) of the "equator." Cut both north and south of the orange, but don't cut across the "equator" yet. **(This will require adult supervision.)**

5. When you have finished making these cuts, use your plastic knife to finish making one cut that goes all the way from "pole" to "pole." Using the dull side of the plastic knife, carefully begin to remove the skin from the orange. Try to keep the whole skin in one piece.

6. Lay the peel down and flatten carefully, allowing the places you cut to separate, as in the diagram.

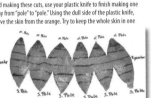

Part B

1. Make a diagram of your flattened orange peel and label the north and south poles, the equator line, the longitudinal lines, the latitudinal lines, and the continents.

2. Compare the orange peel "map" with a flat map of the world. Compare with a globe if one is available.

3. Are the areas close to the poles really as large as they are drawn on a flat map of the world?

4. You only drew six longitudinal lines on your orange map. How many longitudinal lines are usually shown on a world map?

5. The longitudinal lines on a globe get closer together as they get closer to the poles. The latitudinal lines on a globe make smaller and smaller circles as they get closer to the poles. Make careful observations of your flat map of the world.

6. Does the flat map show these changes in the longitudinal and latitudinal lines?

7

OBJECTIVES Students will learn how longitudinal and latitudinal lines divide the earth, and how they can be used to identify places on the earth. Some of the problems with flat maps will be illustrated.

NOTE You can incorporate a little more math by pointing out that there are 360° in a circle. The earth's equator also makes a 360° circle. There are 24 hours in a day, so the earth can be divided into 24 time zones. Divide 360 by 24 to get 15° in each time zone. This will be more visual if you have the students draw a circle and then use a protractor to divide the circle into 24 sections.

Permanent markers work best, but aren't essential.

The Science Stuff

The orange peel is a model of a flat map of the world. Scientific models are used to make comparisons with other things, which may be too small to see or may be difficult to understand. Most students need a little help understanding how all the countries and oceans of the world can be placed on a flat map when the earth is actually round. Let's first look at the basic terms and features by referring to a flat map of the world.

The lines that go up and down and connect the North Pole and the South Pole are called longitudinal lines. Most world maps show 24 longitudinal lines that correspond (more or less) to the 24 times zones.

In order to have a starting point for counting lines on a sphere, the longitudinal line that was chosen to be 0° was the one that passes through the Royal Observatory at Greenwich, England. It is sometimes referred to as the Greenwich Meridian or the Prime Meridian.

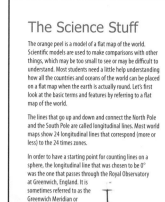

Greenwich, England

8

The equator is a line that circles the earth halfway between the North Pole and the South Pole. Other lines that circle the earth above and below the equator are known as latitudinal lines. The places at and near the equator tend to be hot all year. Places near the North and South Poles tend to be cold all year. Places farther away from the poles and the equator tend to have four seasons — spring, summer, fall, and winter.

Finding where a longitudinal line and a latitudinal line intersect can identify any place on the earth. Maps only label the main lines, but there are many other lines that are not shown on a map. Hurricanes can be tracked if you know the exact longitudinal and latitudinal lines that cross it. Exact positions can be obtained from special satellites in orbit above the earth and transmitted to a Global Positioning System. Ships in the middle of the ocean can identify their positions with a GPS system and radio for help if they need it.

Making Connections

The time zones are not exactly lined up with the longitudinal lines, but each time zone is approximately 15°. The earth is divided into 24 longitudinal lines that are 15° apart. There are four times zones in the United States. If it is 10:00 a.m. in Seattle, Washington (Pacific Time Zone), and you move east (to the right), the time goes up an hour as you cross into another time zone. That means the time in New York City is 1:00 p.m. (Eastern Time Zone). Have you ever wondered why it just couldn't be 10:00 a.m. all over the world at the same time? Most everyone expects 10:00 a.m. to be the time between breakfast and lunch, but if there weren't different time zones, 10:00 a.m. would be in the middle of the night for many countries.

The longitudinal line on the opposite side of the world, 180° from the Greenwich Meridian, is known as the International Date Line. The date changes if you cross this line!

Your orange peel model may look like the earth has many North and South Poles. To avoid the broken look, mapmakers stretch out the northern and southern countries and oceans to be continuous. This causes the countries near the Poles on a flat map of the world to appear distorted. For example, on a flat map, Greenland may appear larger than it does on a globe. Also, on a globe, the distance between any two longitudinal lines is much farther apart at the equator than they are near the Poles. Globes are more accurate than flat maps, but they would not work on the flat pages in a book.

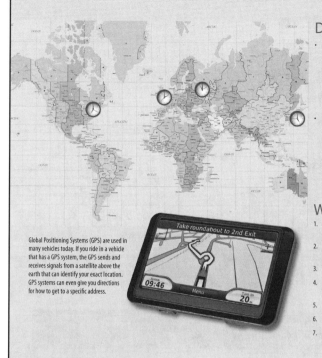

Global Positioning Systems (GPS) are used in many vehicles today. If you ride in a vehicle that has a GPS system, the GPS sends and receives signals from a satellite above the earth that can identify your exact location. GPS systems can even give you directions for how to get to a specific address.

Dig Deeper

- On a map, the Geographic North Pole and the Geographic South Pole are labeled and represent the axis of the earth's spin. The earth also has a Magnetic North Pole and a Magnetic South Pole that are located in different places. You probably won't find the magnetic poles on a regular map. Find where both magnetic poles are located, and show this on a map. Explain what is meant by declination when using a compass.

- There are several Internet sites that can tell you what time it is in other places around the world. Find the time in your hometown. Then find what time it is right now in ten other cities around the world. Record the cities and the times. Now look for a pattern about how the time changes as you go from east to west.

What Did You Learn?

1. Is the International Date Line a longitudinal line or a latitudinal line?

2. What is the name of the starting longitudinal line that is designated as 0°?

3. Which lines go from the North Pole to the South Pole?

4. Which lines circle the earth and are parallel to the equator?

5. What part of the earth doesn't have four seasons?

6. Into how many times zones is the earth divided?

7. What is a GPS device? What can a GPS device in an automobile do?

9

WHAT DID YOU LEARN?

1. Is the International Date Line a longitudinal line or a latitudinal line? *Longitudinal line*

2. What is the name of the starting longitudinal line that is designated as 0°? *The Greenwich Meridian or the Prime Meridian*

3. Which lines go from the North Pole to the South Pole? *Longitudinal lines*

4. Which lines circle the earth and are parallel to the equator? *Latitudinal lines*

5. What part of the earth doesn't have four seasons? *The part of the earth on or near the equator*

6. Into how many times zones is the earth divided? *24*

7. What is a GPS device? What can a GPS device in an automobile do? *It stands for Global Positioning System. It can tell you where you are on the earth. It is usually able to show you how to get to a specific address.*

INVESTIGATION #2

Composition of the Earth

Think about This

Molly laughed at her four-year-old brother, Devin, who was busy digging a hole in their backyard. Molly had been reading a science fiction book called *Journey to the Center of the Earth* by Jules Verne. Molly told Devin that the story in her book was about finding a pathway to the center of the earth and meeting strange creatures on the way. Devin was pretty sure he could take his own journey to the center of the earth by digging a big hole. Have you ever wondered what you would find if you could travel through the center of the earth? What do you think we would find if we dug a really deep hole in the earth? If you make a model of the earth, you'll be able to see what you would find deep inside.

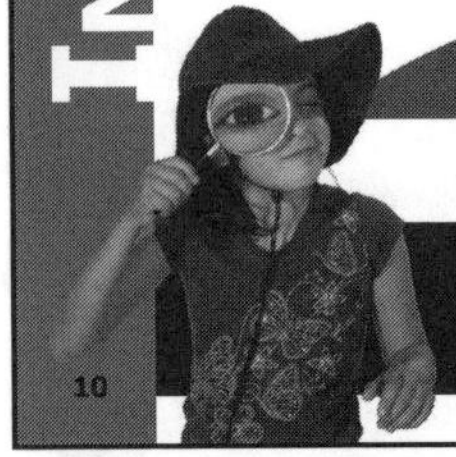

Procedure & Observations

Part A

1. Make a model of the interior of the earth with modeling clay. Shape your model of the earth into half a sphere, like the earth had been cut in half to show what was inside. Use the drawing on page 11 to help make your clay model.

2. Use a different color of clay for each part: crust, mantle, outer core, and inner core. Try to make the sizes of each part realistic.

3. Write the names of the parts that make up the earth on the slips of paper. Use your reference sources to find additional information about each layer. On the back of the slips of paper, write at least two facts that you have found about each layer.

The Investigative Problems
What is inside of the earth?

10

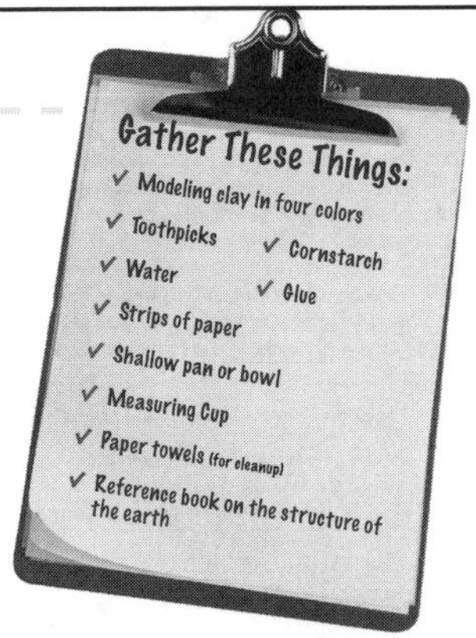

Gather These Things:

- ✓ Modeling clay in four colors
- ✓ Toothpicks
- ✓ Water
- ✓ Strips of paper
- ✓ Shallow pan or bowl
- ✓ Measuring Cup
- ✓ Paper towels (for cleanup)
- ✓ Reference book on the structure of the earth
- ✓ Cornstarch
- ✓ Glue

4. Glue each of the labels to a toothpick, and put the toothpicks in the correct part in the clay model.

5. Put this information in your answer booklet.

6. Look for patterns in the temperature and pressure going from the crust to the inner core of the earth. Write what you find about changes in temperatures and pressures going from the crust to the inner core.

Labels: Mantle, Crust, Outer Core, Inner Core

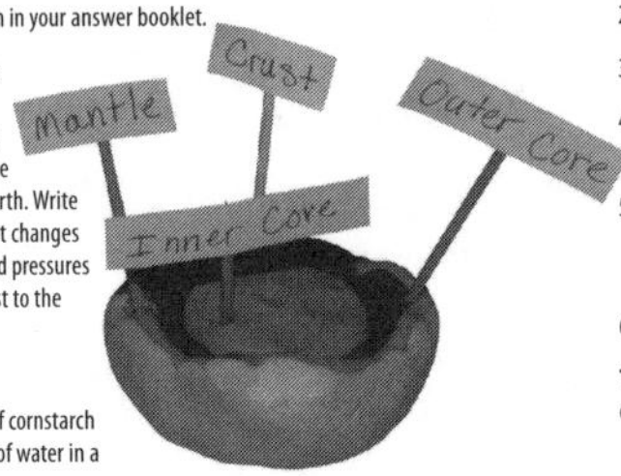

Part B

1. Combine one cup of cornstarch with about ²/₃ cup of water in a shallow pan. Stir for a few minutes until you have an evenly mixed liquid. Then stir in another ¹/₂ cup of cornstarch. Add a little more water if this is too dry.

2. What happens to the mixture as you stir in the additional cornstarch?

3. What happens to the mixture when you stop stirring and let the mixture set?

4. Push your finger through the mixture slowly. When you push your finger through the mixture slowly, does it have properties of a solid or a liquid?

5. Now pick up some of the mixture and make it into a ball. Observe what happens when you squeeze it and when you let it sit in your hand. What happens when you squeeze the ball in your hands?

6. What happens when you stop squeezing it and let it sit in your hand?

The mixture can be poured down the drain, but be sure to run lots of hot water down the drain while you are cleaning up.

11

OBJECTIVES

The earth's crust is the part of the earth most familiar to us, but students will learn that the interior of the earth also consists of areas known as the mantle, outer core, and inner core.

NOTE

Students tend to think the earth's crust is much thicker than the models show. Try to help them realize that in their model of the interior of the earth, the crust will be very thin compared to the rest of the earth's interior.

The Science Stuff

Anyone could easily dig through a layer of topsoil, which is only a few centimeters (one to two inches) thick in some places and is completely eroded away in other places. The soil below the topsoil contains mostly gravel, sand, silt, and clay. This layer may be shallow, or it may be very deep. Eventually, there is a point at which solid rock is encountered. For the first several meters (a little over three feet), the temperature of the soil and rocks gets cooler as one digs deeper. Then the temperature begins to increase. Extremely high temperatures are found in the mantle and the outer core. The highest temperatures are found in the inner core.

Most of the earth's crust is composed of hard granite and basalt rock. The crust includes both dry land and ocean floor. The crust is thicker beneath the land continents and thinner beneath the seafloors. In addition to rocks and soil, the crust also contains water, coal, oil, gas, ores, and mineral deposits. Even though this layer looks gigantic to us, the earth's crust should look very thin on your clay model.

Beneath the crust, there is a thick layer known as the mantle. There is nothing living in or below this depth. The rocks in the upper part of the mantle are very dense and solid. If you were to go down deeper into the mantle, the temperature would get very hot making plastic rock that has both properties of solid and liquid. The mixture of cornstarch and water behaves like plastic rock in the mantle in some ways. This very hot rock is under great pressure, which causes it to behave like a liquid when the pressure is reduced. This layer extends to a depth of about 2,900 kilometers (1,800 miles).

Sometimes the material in the mantle is able to squeeze through a crack in the crust. The hot magma that comes from a volcano originates in the mantle. Convection-type currents in the mantle may affect the movement of the earth's crustal plates.

Beneath the earth's mantle is the outer core, composed mostly of iron in a liquid state. There may also be a little nickel. The temperature is even greater in the core than in the mantle. There may be convection-type currents in this area that play a role in producing the earth's magnetic field.

The inner core is also composed mostly of iron. The temperatures in the inner core may be as high as 4,500°C. The core appears to be solid, because the pressure on the particles is greater than the forces that normally cause hot particles to expand.

The make-up of the earth can be studied by observing how seismic waves from earthquakes travel through the earth. Information coming from various seismic stations around the world allows scientists to determine the speed of these waves and the paths they take.

Crust
Temp: 32°F
62 miles thick
Contains gravel, sand, silt, clay, hard granite, and basalt rock

Mantle
Temp: 1,832°F
1,802 miles thick
Contains magnesium, iron, aluminum, silicon, and oxygen

Outer Core
Temp: 6,692°F
1,367 miles thick
Contains mostly liquid iron and sulfur

Inner Core
Temp: 7,772°F
778 Miles thick
Contains mostly solid iron

Making Connections

Scientists often study drill cores (see image below) of rock that have been brought up, as drillers look for sources of oil deep underground. The deepest oil wells are only a few miles deep. Several years ago, a group of scientists had a goal of drilling even deeper and reaching the Moho, the boundary between the crust and the mantle. A huge amount of money went into this project. The drillers came close, but so far the Moho region hasn't been reached.

Most scientists believe there is a connection between the hot liquids in the earth's core and the earth's magnetic field. The moon does not seem to have hot liquids inside it like the earth does. It also does not have a noticeable magnetic field around it.

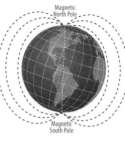

Magnetic North Pole

Magnetic South Pole

Dig Deeper

- The iron in the earth's core is thought to be the source of the earth's magnetic field. What are some of the explanations for how this might occur? (Note: Be sure to look at the rapid decay theory.)

- Do some more research on the Moho project and make a poster of what was found. Summarize what you learned in your own words and present this as an oral or written report.

What Did You Learn?

1. In which layer of the earth are solid rocks found that are not extremely hot?

2. The materials in the lower part of the mantle are extremely hot, but they are thought to be in a plastic state in places. Under what conditions does this plastic behave like a solid?

3. Under what conditions does material from the mantle rise up into the crust or even to the surface of the earth?

4. Most scientists believe the core of the earth is made of what elements?

5. The three main states of matter are solid, liquid, and gas. What is meant by a plastic state?

6. Is the earth's crust thicker under the continents or under the oceans?

7. Some geologists believe there are convection-type currents in parts of the earth's mantle and core. What are some ways these currents might affect the earth?

8. What are some of the natural resources found in the earth's crust?

Drill cores

WHAT DID YOU LEARN?

1. In which layer of the earth are solid rocks found that are not extremely hot? *In the crust*

2. The materials in the lower part of the mantle are extremely hot, but they are thought to be in a plastic state in places. Under what conditions does this plastic behave like a solid? *When they are under great pressure and very high temperatures*

3. Under what conditions does material from the mantle rise up into the crust or even to the surface of the earth? *When there are cracks in the earth's crust*

4. Most scientists believe the core of the earth is made of what elements? *Iron and nickel*

5. The three main states of matter are solid, liquid, and gas. What is meant by a plastic state? *Something in-between a solid and a liquid. It often depends on the amount of pressure and temperature on the material.*

6. Is the earth's crust thicker under the continents or under the oceans? *Under land*

7. Some geologists believe there are convection-type currents in parts of the earth's mantle and core. What are some ways these currents might affect the earth? *They might cause parts of the earth's crust to slowly move. They may play a role in producing the earth's magnetic field.*

8. What are some of the natural resources found in the earth's crust? *In addition to rocks and soil, the crust also contains water, coal, oil, gas, ores, and minerals deposits.*

Why Is Everything Moving?

Think about This
Rico and Tony were looking at a picture in their Bible of the Garden of Eden and trying to guess what the world might have been like then. Rico mentioned that there might have only been one big continent at first. Tony wondered if there were tall mountains and deep oceans on the earth in the beginning. Do you think major landmasses and oceans might have been different in the days of Adam and Eve than they are today?

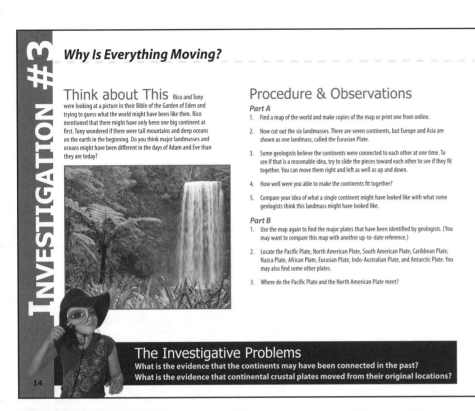

The Investigative Problems
What is the evidence that the continents may have been connected in the past?
What is the evidence that continental crustal plates moved from their original locations?

Procedure & Observations

Part A

1. Find a map of the world and make copies of the map or print one from online.

2. Now cut out the six landmasses. There are seven continents, but Europe and Asia are shown as one landmass, called the Eurasian Plate.

3. Some geologists believe the continents were connected to each other at one time. To see if that is a reasonable idea, try to slide the pieces toward each other to see if they fit together. You can move them right and left as well as up and down.

4. How well were you able to make the continents fit together?

5. Compare your idea of what a single continent might have looked like with what some geologists think this landmass might have looked like.

Part B

1. Use the map again to find the major plates that have been identified by geologists. (You may want to compare this map with another up-to-date reference.)

2. Locate the Pacific Plate, North American Plate, South American Plate, Caribbean Plate, Nazca Plate, African Plate, Eurasian Plate, Indo-Australian Plate, and Antarctic Plate. You may also find some other plates.

3. Where do the Pacific Plate and the North American Plate meet?

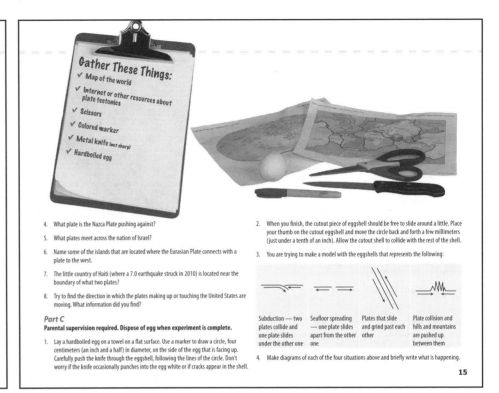

Gather These Things:
- ✓ Map of the world
- ✓ Internet or other resources about plate tectonics
- ✓ Scissors
- ✓ Colored marker
- ✓ Metal knife (not sharp)
- ✓ Hardboiled egg

4. What plate is the Nazca Plate pushing against?

5. What plates meet across the nation of Israel?

6. Name some of the islands that are located where the Eurasian Plate connects with a plate to the west.

7. The little country of Haiti (where a 7.0 earthquake struck in 2010) is located near the boundary of what two plates?

8. Try to find the direction in which the plates making up or touching the United States are moving. What information did you find?

Part C
Parental supervision required. Dispose of egg when experiment is complete.

1. Lay a hardboiled egg on a towel on a flat surface. Use a marker to draw a circle, four centimeters (an inch and a half) in diameter, on the side of the egg that is facing up. Carefully push the knife through the eggshell, following the lines of the circle. Don't worry if the knife occasionally punches into the egg white or if cracks appear in the shell.

2. When you finish, the cutout piece of eggshell should be free to slide around a little. Place your thumb on the cutout eggshell and move the circle back and forth a few millimeters (just under a tenth of an inch). Allow the cutout shell to collide with the rest of the shell.

3. You are trying to make a model with the eggshells that represents the following:

Subduction — two plates collide and one plate slides under the other one	Seafloor spreading — one plate slides apart from the other one	Plates that slide and grind past each other	Plate collision and hills and mountains are pushed up between them

4. Make diagrams of each of the four situations above and briefly write what is happening.

OBJECTIVES
Students will investigate the theory that the earth's crust consists of different plates that are slowly moving. The plates may bump into each other, move apart, or slide past each other. Understanding these movements will introduce other ideas about earthquakes and volcanoes.

NOTE
This lesson introduces the most current theories about plate tectonics, but students should be reminded to keep an open mind when researching some of the ideas in this lesson.

Many references routinely assume that these processes took millions and millions of years. These assumptions will be dealt with in subsequent lessons.

Place a thin sheet of paper over a map and trace the outlines of the continents. Cut the shapes of the continents from this paper.

The Science Stuff

Ideas known as plate tectonics and continental drift are popular among geologists today. Both ideas are logical, so we will investigate them and some of the evidence that supports them. However, this is a good time to remind you of two things: (1) Good scientists always keep an open mind about these ideas and are willing to consider new research and new explanations. (2) Good scientists continue to critically analyze and debate the research. When scientific explanations are thoroughly debated and analyzed, the correct scientific explanations should become stronger and the false ones should be put aside. Keep in mind that some of these ideas may be adjusted or discarded in the future.

The hardboiled egg is a useful (but not perfect) model of the earth. The eggshell represents the crust of the earth, with the broken circle of eggshell representing one of the plates of the crust. The egg white represents the plastic rock of the mantle. The egg yellow represents the earth's core. The boundaries of plates are the most common places for earthquakes and volcanoes to occur.

Scientists have been able to measure the speed and directions of some of the earth's plates. There are four common things that seem to happen when two plates are next to one another.

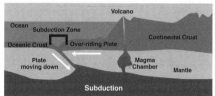

Subduction occurs when two plates collide, and one plate slides under the other one. This most often happens along ocean coasts. The lighter rocks of the continental crust tend to ride over the top of the heavier oceanic crust. This causes the oceanic crust to get pushed down and eventually to melt into the mantle. Both earthquakes and volcanoes are found along subduction zones. This activity can occur off shore or on land.

Plate collision can also occur as plates push against each other but neither slides under the other one. Instead, the plates buckle, forming mountains. The Himalayan Mountains are thought to have formed in this way.

Plate sliding occurs as plates slide against each other. One of the most famous places where this is happening is known as the San Andreas Fault in western California, which has been the source of many earthquakes over the years. (There is actually a system of faults along the main fault.) The Pacific Plate lies to the west of the San Andreas Fault, and the North American Plate lies to the east. The Pacific Plate moves northwest for a distance of about five centimeters (just over an inch and a half) each year, grinding and sliding past the North American Plate. Sometimes the two plates lock up, but the pressure on the plates continues to build up. If the plates suddenly break loose from their locked positions, there is an earthquake.

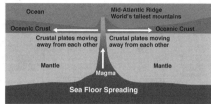

Another kind of crustal plate movement is known as sea floor spreading. Instead of two plates pushing on each other, they are moving away from each other. This seems to be what is happening along the Mid-Atlantic Ridge. As the plates move away from each other, hot magma rises up into the crack between the plates and eventually reaches the ocean floor. It then begins to cool and harden, forming new ocean crust material and producing some of the world's tallest mountains.

North America and Europe, as well as South America and Africa, may have been joined together at one time. The Mid-Atlantic Ridge could have been where they were joined originally.

The name Pangaea is given to a hypothetical original landmass. This idea seems logical because the continents seem to fit together like a giant jigsaw puzzle. Evolutionary scientists believe the continents have been slowly moving apart at the same rate for millions of years. Some creation scientists believe continents separated in the past, but this occurred quickly as a result of the Genesis Flood. Some creation scientists are skeptical of the whole idea of Pangaea.

16

Making Connections

Actually, the explanations about Pangaea and plate tectonics have parts that many evolutionists and creationists agree upon, especially the idea that crustal plates are slowly moving. This movement seems to be a cause for both earthquakes and volcanoes.

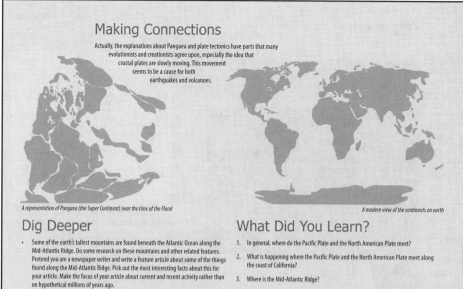

A representation of Pangaea (the Super Continent) near the time of the Flood

A modern view of the continents on earth

Dig Deeper

- Some of the earth's tallest mountains are found beneath the Atlantic Ocean along the Mid-Atlantic Ridge. Do some research on these mountains and other related features. Pretend you are a newspaper writer and write a feature article about some of the things found along the Mid-Atlantic Ridge. Pick out the most interesting facts about this for your article. Make the focus of your article about current and recent activity rather than on hypothetical millions of years ago.

- The Mount St. Helens volcanic eruption in 1980 is thought to have been caused by the movement of two crustal plates. Do some research on what geologists think may have caused Mount St. Helens to become an active volcano. Find where there have been other volcanoes and earthquakes in this area.

What Did You Learn?

1. In general, where do the Pacific Plate and the North American Plate meet?

2. What is happening where the Pacific Plate and the North American Plate meet along the coast of California?

3. Where is the Mid-Atlantic Ridge?

4. Is the Mid-Atlantic Ridge a region of subduction or seafloor spreading?

5. What happens frequently along or near the San Andreas Fault in western California?

6. The Himalayan Mountains are thought to have formed when two crustal plates did what?

7. What do geologists believe about a landmass known as Pangaea?

17

WHAT DID YOU LEARN?

1. In general, where do the Pacific Plate and the North American Plate meet? *Along the western coastline of the United States and Canada*

2. What is happening where the Pacific Plate and the North American Plate meet along the coast of California? *The two plates are sliding and grinding past each other as the Pacific Plate moves northward at the rate of about 5 centimeters (just over an inch and a half) per year.*

3. Where is the Mid-Atlantic Ridge? *It runs north and south between North America and Europe and between South America and Africa.*

4. Is the Mid-Atlantic Ridge a region of subduction or seafloor spreading? *Seafloor spreading.*

5. What happens frequently along or near the San Andreas Fault in western California? *There are frequent earthquakes there.*

6. The Himalayan Mountains are thought to have formed when two crustal plates did what? *Collided into each other, but neither plate slid under the other one.*

7. What do geologists believe about a land mass known as Pangaea? *Pangaea may have been the original land mass that broke apart and formed today's continents.*

Earthquake

Think about This

It was October 17 in San Francisco's Candlestick Park. The introductions of the San Francisco Giants and the Oakland Athletics players began promptly at 5:00 p.m., as the third game of the 1989 World Series was about to begin. At 5:04, a 6.9 earthquake shook the stadium. The press box swayed, the upper floor moved up and down like ocean waves, and power was lost. After several seconds, the shaking stopped. The stunned fans in the stadium and the millions who were watching this on TV or listening on the radio were relieved to realize that none of the park structures had fallen on the fans. No one inside the park was hurt, but outside was a different story. A number of bridge sections and buildings in and around San Francisco had collapsed, causing several deaths and widespread damage. Do you know that the shaking ground during an earthquake can literally cause buildings and other structures to fall? What is an earthquake?

Procedure & Observations

Part A
Wear safety glasses or goggles as you do these investigations.

1. Take a Popsicle stick and hold it at both ends, using two hands. Slowly bend the stick, making an arch. Release the pressure on the stick and observe.

2. Describe what happens when the pressure on the stick is released.

3. Hold and bend the stick as you did before, but keep bending the ends toward each other until it breaks.

4. At what point did the stick break?

The Investigative Problems

What causes an earthquake?
Are some earthquakes more powerful than others?

18

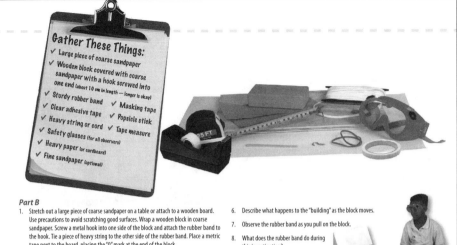

✓ Large piece of coarse sandpaper
✓ Wooden block covered with coarse sandpaper with a hook screwed into one end (about 10 cm in length — longer is okay)
✓ Sturdy rubber band   ✓ Masking tape
✓ Clear adhesive tape   ✓ Popsicle stick
✓ Heavy string or cord   ✓ Tape measure
✓ Safety glasses (for all observers)
✓ Heavy paper (or cardboard)
✓ Fine sandpaper (optional)

Part B

1. Stretch out a large piece of coarse sandpaper on a table or attach to a wooden board. Use precautions to avoid scratching good surfaces. Wrap a wooden block in coarse sandpaper. Screw a metal hook into one side of the block and attach the rubber band to the hook. Tie a piece of heavy string to the other side of the rubber band. Place a metric tape next to the board, placing the "0" mark at the end of the block.

2. Construct a model building out of heavy paper and tape. Make it about 25–30 centimeters (just under a foot) high and attach it securely to the block with masking tape.

3. Now begin to slowly and steadily pull the block across the sandpaper. The rubber band will begin to stretch, but continue to exert pressure with a slow, steady pull until the block moves. **(Caution: Observers must wear safety glasses and not get too close to the block, as it may move in an unpredictable way.)**

4. Write your observations about how the block moves. Notice if it moves forward smoothly or if it surges forward suddenly.

5. Measure the distance the block moves forward in centimeters.

6. Describe what happens to the "building" as the block moves.

7. Observe the rubber band as you pull on the block.

8. What does the rubber band do during this investigation?

9. Stretch the rubber band around two stationary objects. Pull back on one side and then quickly release it.

10. Does the rubber band continue to vibrate for a few seconds?

Optional

1. Repeat this investigation, but use fine sandpaper this time instead of coarse sandpaper.

2. Measure the distance the block moves forward in centimeters.

3. Compare what happened this time with what happened when using coarse sandpaper.

19

OBJECTIVES

Students will recognize that most of the world's earthquakes tend to occur near where crustal plates meet or near other fault systems. They will understand that earthquakes can be explained by the elastic rebound theory.

NOTE

There are no exact dimensions for the size of the wooden block, but don't use one that is too small. There are often pieces of scrap lumber that are available. Everyone observing this should wear safety glasses, because the block could surge into someone who happens to be too close.

WHAT DID YOU LEARN?

1. According to the elastic rebound theory, how do the rocks break loose from a position of tension – do they suddenly surge forward or do they gradually move forward? *They tend to suddenly surge forward.*

2. According to the elastic rebound theory, what often causes the ground to shake during an earthquake? *The rocks are bent and are under tension until they break loose. There is a brief period when the rocks vibrate and shake until they reach a state that is free from tension.*

3. An earthquake begins as locked-up sections of rocks break free. Is the tension on the rocks increased or decreased after an earthquake? *The tension is decreased.*

4. Sometimes buildings fall in during an earthquake. What are some of the things that play a big role in how well a building can withstand an earthquake? *The engineering design and the materials used in buildings*

The Science Stuff

The simple models used in this investigation will help you understand some of the causes of earthquakes. You should be aware that these models are not perfect examples of what happens, but they are helpful learning aids and provide some basic explanations for what happens during an earthquake.

The Popsicle activity can help you understand what happens when pressure is applied to rocks from different places. When this kind of pressure is applied to a section of rocks, the rocks may begin to bend. The pressure on the rocks may cause one section to suddenly slip so that the tension is released or the rocks may crack and break. Large cracks in rocks that have broken in this way are called faults. Large faults are often places where earthquakes begin.

Pulling the block over sandpaper can help you understand the elastic rebound theory of earthquakes. What you most likely observed was that the block surged forward suddenly when it began to move. This movement caused the "building" attached to the block to shake back and forth. This is similar to what happens when one section of rocks (or an entire crustal plate) is moving and rubbing against another one.

Sometimes the rocks get stuck together when there are movements in the crust. The tension on the rocks will continue to build, just as the tension on the rubber band increased. The rocks can only bend so much until they finally slip loose or break. There is a sudden slippage of the rocks as the rocks break free and rebound to a place where the tension is released. When this happens, there is an earthquake. The strength of the earthquake depends on how much tension had built up when the rocks breaks free. After the tension is released, it can build up again if the plates continue to move.

You know that guitar strings and rubber bands continue to vibrate for a few seconds after you pull back on them and then quickly release them. The earth also tends to keep on vibrating back and forth for a short time after a rock section suddenly surges forward to a new place.

The shaking of the earth during an earthquake also shakes buildings and other things on the earth. The engineering design and the materials used in buildings play a big role in how well a building can withstand an earthquake.

As you already know, earthquakes are most likely to occur where one plate in the earth's crust is pushing against another plate or where two plates are sliding past each other. For example, there are many earthquakes near the "Ring of Fire," which is where the Pacific Plate meets other crustal plates.

Evidence of elastic rebound in a plowed field after a magnitude 6.5 earthquake

Moving plate

Contact of plates

Moving plate

Earthquakes along the San Andreas Fault give us a good idea of what happens as crustal plates slowly move. The Pacific Plate is slowly moving past the North American Plate. From time to time, the rocks in these two plates get stuck on each other. As the plate continues to move, the locked-together rock sections begin to bend from the stress on them. Eventually the rocks break loose, rebounding back into a place where the tension on the rocks decreases. The rebounding, vibrating rock sets up a series of violent vibrations that can shake a large area for several seconds.

Making Connections

One of the strongest earthquakes in U.S. history hit New Madrid, Missouri, in 1811. There were no instruments to measure it, but its effects were huge and were even felt in Massachusetts. Fortunately there were only a few people who lived in the area. The New Madrid fault system continues to produce small earthquakes in several surrounding states, but is not in one of the earth's two main earthquake belts. These faults are not at the boundary of two plates but are deep cracks beneath the surface.

Dig Deeper

- What is a tsunami? What causes a tsunami to form? Tell about the tsunami warning system that is in place in most countries. Scientists hope to be able to learn enough about what causes earthquakes to be able to predict when one is about to happen. Try to find more information about scientific efforts to predict earthquakes and tsunamis.

- Do an Internet search for video clips inside buildings during an earthquake. For example, YouTube and other sources have security videos of buildings being filmed when an earthquake struck. Describe in detail what happened to the building and the things inside the building as the earthquake shook the area. Some of the clips show the time frame. Notice how long the earthquake lasts.

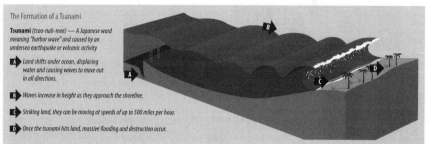

The Formation of a Tsunami

Tsunami *(tsoo-nah-mee)* — A Japanese word meaning "harbor wave" and caused by an undersea earthquake or volcanic activity

A Land shifts under ocean, displacing water and causing waves to move out in all directions.

B Waves increase in height as they approach the shoreline.

C Striking land, they can be moving at speeds of up to 500 miles per hour.

D Once the tsunami hits land, massive flooding and destruction occur.

What Did You Learn?

1. According to the elastic rebound theory, how do the rocks break loose from a position of tension — do they suddenly surge forward or do they gradually move forward?

2. According to the elastic rebound theory, what often causes the ground to shake during an earthquake?

3. An earthquake begins as locked-up sections of rocks break free. Is the tension on the rocks increased or decreased after an earthquake?

4. Sometimes buildings fall in during an earthquake. What are some of the things that play a big role in how well a building can withstand an earthquake?

5. Where are earthquakes most prone to occur?

6. What kind of crustal plate movement is occurring along the San Andreas Fault?

7. Is the New Madrid fault system located at the boundary of two major crustal plates?

8. One of the main earthquake belts in the earth is known as the "Ring of Fire." What seems to be happening all along the coastline of the Pacific Ocean to cause so many earthquakes?

---

5. **Where are earthquakes most prone to occur?** *Earthquakes tend to occur along the boundary of crustal plates or along other fault lines, especially where one plate in the earth's crust is pushing against another plate or where two plates are sliding past each other.*

6. **What kind of crustal plate movement is occurring along the San Andreas Fault?** *The Pacific Plate is slowly moving past the North American Plate.*

7. **Is the New Madrid fault system located at the boundary of two major crustal plates?** *No. These faults are not at the boundary of two plates, but are deep cracks in the crust beneath the surface.*

8. **One of the main earthquake belts in the earth is known as the "Ring of Fire." What seems to be happening all along the coastline of the Pacific Ocean to cause so many earthquakes?** *The Pacific Plate seems to be bumping into or sliding past other plates.*

Living with Earthquakes

Think about This
On December 16, 2003, a 6.6 magnitude earthquake struck an area in southeastern Iran, killing 31,000 people. On October 15, 2006, a 6.7 earthquake struck the island of Hawaii. There were no casualties. On September 11, 2008, a 6.8 earthquake hit Hokkaido, Japan. There were no casualties.

All three areas were hit by earthquakes of similar strength. Why do you think one of these earthquakes was much more deadly than the other two?

Damaged city street in Mianyang, Sichuan, China

This chart gives information about some of the world's biggest earthquakes since 2001.

Date	Location	Deaths from	Richter Scale
June 2001	Coastal S. Peru	75	8.4
Nov. 2002	Denali Nat. Park, Alaska	0	7.9
Sept. 2003	SE of Japan	0	8.3
Dec. 2004	SW of New Zealand	0	8.1
Dec. 2004	W. Sumatra, Indonesia	229,000	9.3
March 2005	Nias region, Sumatra	1,303	8.6
May 2006	Tonga	0	8.0
Nov. 2006	Kuril Islands, Russia	0	8.3
April 2007	Solomon Islands	52	8.1
Sept. 2007	Sumatra, Indonesia	25	8.5
May 2008	Sichuan Province, China	69,197	7.9
Sept. 2009	Samoa Islands	189	8.0
Jan. 2010	Haiti	230,000	7.0
Feb. 2010	Chile	400	8.8
April 2010	Qinghai Province, China	2,039	6.9

The Haitian earthquake set in motion the conditions for over 50 aftershocks over 4.5 or higher. Nearly 1,000,000 people became homeless from the effects of the quake, which included damage or destruction of hospitals, communications systems, and electrical networks.

The Investigative Problems
How can buildings and other structures be designed to better withstand earthquakes? Where are the most earthquake-prone parts of the earth?

22

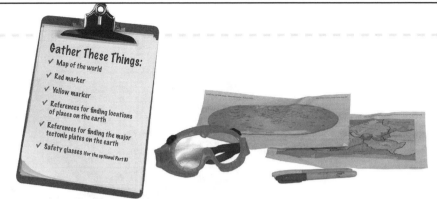

Gather These Things:
✓ Map of the world
✓ Red marker
✓ Yellow marker
✓ References for finding locations of places on the earth
✓ References for finding the major tectonic plates on the earth
✓ Safety glasses (for the optional Part B)

Procedure & Observations

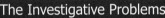

Part A
1. Get a copy of a world map you can mark on (from your teacher). Use the chart on the previous page to find information about recent major earthquakes. Use references to help you find places you don't know.

2. Mark the earthquakes listed on page 22 by making a small red circle at or near each location on the map. Next to the red dot, write the date of the earthquake.

3. When you finish, look for patterns on the map about where most recent earthquakes have occurred. What patterns did you find?

4. Now look in a reference source to find the major tectonic plates on the earth. About eight major plates have been identified, along with a few other smaller plates.

5. Do you see a connection between earthquakes on your map and where the plates meet?

6. About how many of the earthquakes on your map were located where two plates meet?

7. Color the regions with the most recent strong earthquakes yellow.

8. Do strong earthquakes always cause a large number of deaths?

Part B (Optional)
1. Build two houses of toothpicks and gumdrops. Make them as much alike as you can. Tape both houses securely to a wooden board. Put the wooden board holding one house directly on another board and hold in place with sturdy rubber bands.

2. For the other house, put marshmallows or marbles between the two wooden boards and hold the boards in place with sturdy rubber bands. Place a rock on top of each house. Make sure the house can hold up the weight of the rock before you begin the shaking.

3. Shake each house back and forth for several seconds. Move your hand three centimeters (a little over an inch) forward and three centimeters back as you shake the board. Shake the houses the same way each time.

4. Explain what happens to each of the two houses after they have been shaken.

5. Do you think the marshmallows (or marbles) provided a way to make the house less likely to fall in?

23

OBJECTIVES
Students will look for patterns in the locations of earthquakes around the world. They will learn some of the things that can be done to minimize earthquake damage in these areas.

NOTE
This is a good opportunity to integrate a science lesson with a geography/map lesson. Be sure to have to some good maps of the world. Up-to-date atlases and Internet sites should enable students to find each of the locations where major earthquakes have occurred. The "Ring of Fire" is a pattern students can identify. You might want to make a list of other geographical locations in the world and have the students predict whether they think these places are prone to earthquakes or not. Before doing the investigations, ask students if they have any ideas about what people in earthquake-prone areas might do to make homes and buildings safer.

Teachers should be careful to help students understand the dangers of earthquakes without instilling excessive fears. Students should learn what to do in the event of an earthquake and know where to go to be safest. They should be aware that in the United States and in many other countries, building codes help make buildings and other structures much safer. An example of a major earthquake in a country that did not have required building codes was the 2010 Haiti earthquake. Thousands of people died in this quake, because buildings fell in on them. The number of deaths would have been much less if the builders had used better designs and materials.

Today geologists estimate the energy that is released by an earthquake by using the Moment Magnitude Scale. The numbers mentioned on the news are often MMS numbers. The two scales have many similarities, but the Richter Scale continues to be the one that most people are familiar with.

The Science Stuff

Note that the strength of the earthquakes is given as a Richter scale number that ranges from 0 to 10. This is a scale of numbers based on the nature of the seismic waves an earthquake produces. The larger numbers indicate the stronger earthquakes. Earthquakes are studied with a device known as a seismograph.

Small earthquakes with Richter scale numbers less than 3 occur every day in many places over the earth. They don't usually cause major damage, although they may be felt as a minor shaking. The larger earthquakes (5.5 or greater) are more dangerous and fortunately are also more rare. Earthquakes that have magnitudes of 8 and 9 are very rare. They are extremely powerful, but even these do not always cause more deaths and destruction than smaller ones. Safe building standards keep many houses and building from falling in on people inside.

Earthquakes occur frequently in places like Japan, Hawaii, and California, but the death rate is seldom high. All three places have large populations. There are also many tall buildings there. The main reason for the lower death rate in these places is in the design, construction, and selection of building materials for the houses, schools, offices, and other buildings.

The kind of building material is an important thing to consider. For example, buildings made from bricks that are reinforced with steel beams are less likely to fall than ones without steel beams. In order to make buildings safer during an earthquake, they should be built both strong and flexible (able to twist and bend without breaking).

Another thing to consider is the building design. Engineers have found ways to design houses and buildings that can withstand the effects of large earthquakes. Some kinds of design prevent the structure from absorbing all of the energy from the earthquake. If you did the optional investigation, you probably found that a building isolated somewhat from the foundation had less shaking than a building that was fixed securely to the foundation. Marshmallows or marbles between the two boards provided a way for the house to only move slightly on its foundation. This is somewhat like shock-absorbing springs in a car that give you a smoother ride.

In places where there are many deaths from an earthquake, the reason can usually be traced to houses and buildings that fall on the people inside. In the United States, buildings must meet building codes that require certain safety features to be present. In many places around the world, there are no building or safety codes that must be met.

Some of the strongest earthquakes do not always cause a large number of deaths. The death rate during an earthquake could be greatly reduced if safer building standards were enforced.

There are two main earthquake belts, although earthquakes can also occur in other places. One belt follows the coastline around the Pacific Ocean and is known as the "Ring of Fire." The other belt is next to the Mediterranean Sea and extends to southern Asia. Most earthquakes in this belt are at or near where there is contact between the European and African plates and the Arabian and Indian plates.

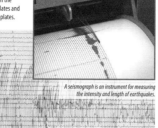

A seismograph is an instrument for measuring the intensity and length of earthquakes.

Making Connections

There have been a number of major earthquakes in the history of the United States in addition to the 1811 New Madrid earthquake. In 1906, an earthquake that measured 8.3 on the Richter scale hit San Francisco. The quake and an ensuing fire destroyed much of the city. In 1964, a violent 8.5 earthquake hit Alaska and destroyed downtown Anchorage. Both of these earthquakes were in the Pacific earthquake belt.

During a strong earthquake, loose, moist soil can be shaken in such a way that the soil turns into liquid mud. The process, known as liquefaction, can cause large cracks to form in the ground, buildings to sink and pull apart, and landslides to occur.

Dig Deeper

- What does the Richter scale for measuring the strength of an earthquake mean? How much stronger is a magnitude of 8 than a magnitude of 7? List several examples of major earthquakes on Earth. Give their strength on the Richter scale, their location, and the date of each quake. Are most of these quakes within one of the earth's two main earthquake belts?

- One of the most destructive earthquakes in history was the January 2010 quake that struck the little country of Haiti. Do some research on what might have caused the quake and what could be done to prevent the kind of damage that occurred there in the future.

- What are some things you should do if you are in a building and an earthquake strikes?

- Dams that hold back large amounts of water must also be designed to withstand earthquakes. Explain why dams are not usually built in earthquake-prone areas. If they are, what safety precautions are taken to make them better able to withstand an earthquake?

What Did You Learn?

1. The strength of an earthquake is reported as a number from 0 to 10. What is this scale of numbers called?

2. What is the name of the instrument that is used to study and identify earthquakes?

3. About how often do earthquakes occur throughout the world — every day or about once a month?

4. Which earthquake is more powerful — one that measures 3 on the Richter scale or one that measures 9 on the Richter scale?

5. What is the advantage of designing a building that moves slightly on its foundation?

6. What are some of the major earthquakes that have struck the United States?

7. What are some ways the government of a country could reduce the deaths and damage caused by an earthquake?

8. Where are the two main earthquake belts in the earth?

Pause and Think

Earthquakes are often mentioned in Scripture, as a part of history, as well as a part of prophecy. For example, there was an earthquake when Jesus was crucified and another one on the first Easter morning as an angel rolled away the stone to Jesus' grave (Matthew 28:2). Once Paul and Silas were in prison in the district of Macedonia when a violent earthquake shook the foundations of the prison so that all the doors of the prison were opened (Acts 16:26).

Earthquakes are also prophesied in Scripture. Jesus told His disciples, "There will be earthquakes in various places, and famines. These are the beginning of birth pains" (Mark 13:8).

"On that day his feet will stand on the Mount of Olives, east of Jerusalem, and the Mount of Olives will be split in two from east to west, forming a great valley, with half of the mountain moving north and half moving south. . . . You will flee as you fled from the earthquake in the days of Uzziah king of Judah" (Zechariah 14:4–5).

If you look at the map showing major plates on the earth, you can easily see that two plates meet in Israel. Seismologists have also identified a system of fault lines throughout Israel and nearby countries.

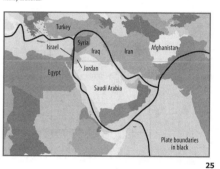

Plate boundaries in black

WHAT DID YOU LEARN?

1. The strength of an earthquake is reported as a number from 0 to 10. What is this scale of numbers called? *The Richter scale*

2. What is the name of the instrument that is used to study and identify earthquakes? *Seismograph*

3. About how often do earthquakes occur throughout the world — every day or about once a month? *Every day*

4. Which earthquake is more powerful — one that measures 3 on the Richter scale or one that measures 9 on the Richter scale? *One that measures 9 on the Richter scale*

5. What is the advantage of designing a building that moves slightly on its foundation? *Structures designed in this way do not absorb as much of the energy from the earthquake as structures that are tightly attached to the earth's foundation.*

6. What are some of the major earthquakes that have struck the United States? *New Madrid earthquake in 1811; San Francisco earthquake in 1906; Anchorage, Alaska, earthquake in 1964. (There are others also.)*

7. What are some ways the government of a country could reduce the deaths and damage caused by an earthquake? *They could enforce building codes for houses, schools, and other buildings that reduce the chances of the buildings falling in during an earthquake. Even if everyone could not afford to do this, the government could be sure that certain buildings, such as schools and hospitals, were built according to safe building codes.*

8. Where are the two main earthquake belts in the earth? *One belt follows the coastline around the Pacific Ocean and is known as the "Ring of Fire." The other belt is next to the Mediterranean Sea and extends to southern Asia.*

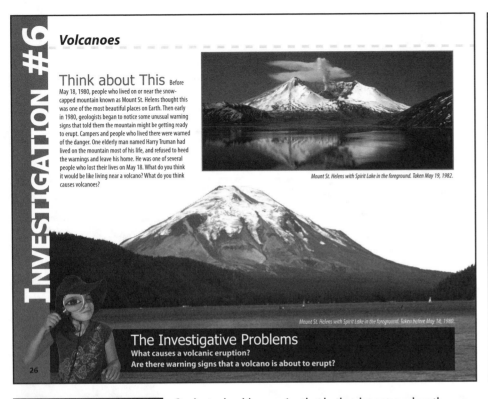

Volcanoes

Think about This

Before May 18, 1980, people who lived on or near the snow-capped mountain known as Mount St. Helens thought this was one of the most beautiful places on Earth. Then early in 1980, geologists began to notice some unusual warning signs that told them the mountain might be getting ready to erupt. Campers and people who lived there were warned of the danger. One elderly man named Harry Truman had lived on the mountain most of his life, and refused to heed the warnings and leave his home. He was one of several people who lost their lives on May 18. What do you think it would be like living near a volcano? What do you think causes volcanoes?

Mount St. Helens with Spirit Lake in the foreground. Taken May 19, 1982.

Mount St. Helens with Spirit Lake in the foreground. Taken before May 18, 1980.

The Investigative Problems

What causes a volcanic eruption?
Are there warning signs that a volcano is about to erupt?

26

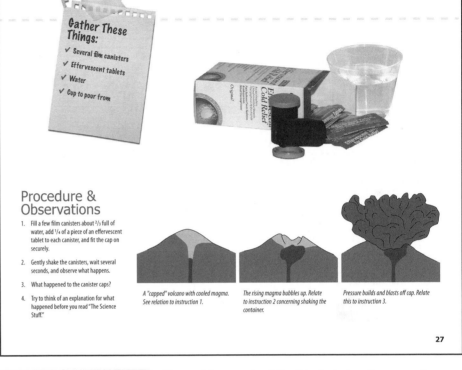

Gather These Things:

✓ Several film canisters
✓ Effervescent tablets
✓ Water
✓ Cup to pour from

Procedure & Observations

1. Fill a few film canisters about ²/₃ full of water, add ¹/₄ of a piece of an effervescent tablet to each canister, and fit the cap on securely.

2. Gently shake the canisters, wait several seconds, and observe what happens.

3. What happened to the canister caps?

4. Try to think of an explanation for what happened before you read "The Science Stuff."

A "capped" volcano with cooled magma. See relation to instruction 1.

The rising magma bubbles up. Relate to instruction 2 concerning shaking the container.

Pressure builds and blasts off cap. Relate this to instruction 3.

27

OBJECTIVES

Students should recognize that both volcanoes and earthquakes tend to occur along crustal plate boundaries, with some exceptions. They should relate what they have learned about magma in the earth's mantle to volcanoes.

NOTE

Film canisters may be difficult to find, since fewer people are using them now. They can still be ordered from science supply houses. An alternative would be to find small plastic containers with lids that don't fit too tightly.

The Science Stuff

Magma originates from the earth's mantle. Because it is less dense than the surrounding rock, it tends to rise upward and flow through cracks in the earth's crust in the form of a hot liquid.

Along with the hot magma, volcanoes often release hot water and gases. Mountains may appear to be smoking as the steam (from the water) escapes from vents in the mountain. Inside the earth, the magma, the water vapor, and other gases exert great pressure on the interior walls of the volcano. If the pressure continues to increase, the gases will push through the containing walls of the volcano. This is similar to what happened in the film canisters when carbon dioxide gas formed and built up pressure inside the canister. Eventually the pressure inside the canister became greater than the forces holding the lid in place, and the lid was forcefully pushed out.

When a volcano suddenly erupts because of a build-up of steam and other gases, volcanic materials and rocks are ejected from the mountain with great force. Volcanic ash and rocks have been known to be ejected with so much force that they are shot several miles into the air. When Mount St. Helens erupted on May 18, 1980, the explosions were the same as 33,000 atomic bomb explosions.

Magma may also break through and flow out onto the surface of the earth. When magma reaches the surface of the earth, it is known as lava. Lava flows are not always released explosively. The active volcanoes in Hawaii periodically release rivers of hot molten lava that flow down the mountains. They can be very devastating to anything in their path. However, the sight of rivers of red, hot lava flowing down a mountain is so spectacular that many tourists get as close to it as they safely can.

Volcanoes, like earthquakes, tend to be found near where large crustal plates meet. However, the Hawaiian Islands are not located where two plates meet. These places are called "hot spots." They are probably related to features under the ocean that allow magma to rise to the surface.

28

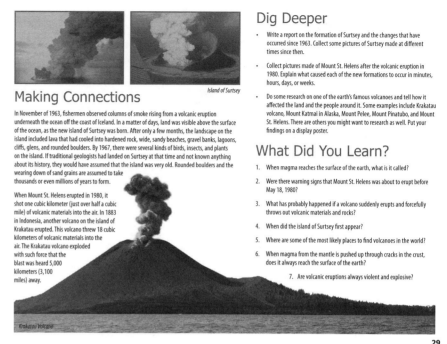

Island of Surtsey

Making Connections

In November of 1963, fishermen observed columns of smoke rising from a volcanic eruption underneath the ocean off the coast of Iceland. In a matter of days, land was visible above the surface of the ocean, as the new island of Surtsey was born. After only a few months, the landscape on the island included lava that had cooled into hardened rock, wide, sandy beaches, gravel banks, lagoons, cliffs, glens, and rounded boulders. By 1967, there were several kinds of birds, insects, and plants on the island. If traditional geologists had landed on Surtsey at that time and not known anything about its history, they would have assumed that the island was very old. Rounded boulders and the wearing down of sand grains are assumed to take thousands or even millions of years to form.

When Mount St. Helens erupted in 1980, it shot one cubic kilometer (just over half a cubic mile) of volcanic materials into the air. In 1883 in Indonesia, another volcano on the island of Krakatau erupted. This volcano threw 18 cubic kilometers of volcanic materials into the air. The Krakatau volcano exploded with such force that the blast was heard 5,000 kilometers (3,100 miles) away.

Krakatau Volcano

29

Dig Deeper

- Write a report on the formation of Surtsey and the changes that have occurred since 1963. Collect some pictures of Surtsey made at different times since then.

- Collect pictures made of Mount St. Helens after the volcanic eruption in 1980. Explain what caused each of the new formations to occur in minutes, hours, days, or weeks.

- Do some research on one of the earth's famous volcanoes and tell how it affected the land and the people around it. Some examples include Krakatau volcano, Mount Katmai in Alaska, Mount Pelee, Mount Pinatubo, and Mount St. Helens. There are others you might want to research as well. Put your findings on a display poster.

What Did You Learn?

1. When magma reaches the surface of the earth, what is it called?

2. Were there warning signs that Mount St. Helens was about to erupt before May 18, 1980?

3. What has probably happened if a volcano suddenly erupts and forcefully throws out volcanic materials and rocks?

4. When did the island of Surtsey first appear?

5. Where are some of the most likely places to find volcanoes in the world?

6. When magma from the mantle is pushed up through cracks in the crust, does it always reach the surface of the earth?

7. Are volcanic eruptions always violent and explosive?

WHAT DID YOU LEARN?

1. When magma reaches the surface of the earth, what is it called? *Lava*

2. Were there warning signs that Mount St. Helens was about to erupt before May 18, 1980? *Yes, and people had been warned to get off the mountain.*

3. What has probably happened if a volcano suddenly erupts and forcefully throws out volcanic materials and rocks? *Steam and other gases probably built up inside the mountain until the pressure became great enough to push through a containing wall.*

4. When did the island of Surtsey first appear? *In 1963*

5. Where are some of the most likely places to find volcanoes in the world? *Volcanoes, like earthquakes, tend to be found where large crustal plates meet. Sometimes they are found in areas known as "hot spots."*

6. When magma from the mantle is pushed up through cracks in the crust, does it always reach the surface of the earth? *No, sometimes magma hardens underneath the surface of the earth.*

7. Are volcanic eruptions always violent and explosive? *No.*

Lessons Learned from Mount St. Helens

One of the most influential secular books written is *Principles of Geology* by Sir Charles Lyell, based on an idea of James Hutton, who said that the same natural processes we observe today occurred in the past. This idea is known as naturalism or uniformitarianism.

Isaac Newton was one of the first scientists to recognize that natural laws, such as the law of gravity, could be applied to the earth, the moon, and every other body in the universe. He recognized that gravity has been operating the same way since the world began.

Hutton and Lyell agreed with Newton and said that natural processes, such as erosion, weathering, sedimentation, deposition, evaporation, condensation, oxidation, and fossilization, operate today in the same way as they did in the past. However, they further claimed that the slow rates at which these processes occur today are the same as in the past. Making logical assumptions is part of the nature of science. However, one should realize that when there is new evidence, these assumptions have to be re-examined.

There is evidence that many landforms formed rapidly as a result of a series of catastrophic events that ranged from minutes to hundreds of years rather than millions of years. Let's focus on one important event as an example of why you should be skeptical of some assumptions. This event is the Mount St. Helens volcanic eruption on May 18, 1980. We will even use Hutton's advice of looking at present-day events to interpret the past.

Assumption 1: Stratified layers are very old. Whenever stratified layers are found in exposed road cuts and other places, it is often assumed that each layer was gradually deposited and that a thick layer of stratified rock is millions of years old.

Observation: If you visit Mount St. Helens today, you will find cliffs where mud, silt, and sand were laid down rapidly in horizontal layers that have since hardened into solid rock. In the picture above, you can see three sections that formed on three different days. The bottom of the cliff is made up of a thick layer of volcanic ash, which was laid down shortly after the volcano first erupted on May 18, 1980. The top layer of rock formed from a mudflow that occurred on March 19, 1982. The middle layer is what we want to focus on. It is made up of numerous layers (or strata) of rock that were laid down on June 12, 1980, in about three hours.

These layers formed as the volcano released large amounts of ground-hugging steam. This steam mixed with volcanic ash and flowed across the ground like a giant river. It often traveled at speeds greater than 100 miles per hour. In only a few hours, sediment from this mudflow had been deposited in horizontal stratified layers. At first the layers were somewhat like wet cement, but eventually the sediment hardened and formed stratified rock.

Assumption 2: Large, complex gully systems are very old. Large gullys, especially the ones like the Grand Canyon, are also assumed to have formed over long periods of time from erosion by a river.

Observation: Many geologists were quite surprised to find that a large, complex gully system formed rapidly in solid ground near Mount St. Helens. During the initial blasts that occurred on May 18, large boulders were torn from the mountain and hurled down the mountain. These boulders made deep scratches in the rock, making them easier to erode. Several steam pit explosions produced large holes in the ground. Later, highspeed, powerful mudflows, and water flows further eroded the rock. These kinds of mudflows resulted in cavitation and hydraulic plucking, processes that further broke up the hard ground and caused gullys to form rapidly.

Top photo: The Upper Muddy River, which was formed from erosion after the eruption of Mount St. Helens. Photograph taken October 10, 1981.

Outline of Mount St. Helens before the eruption

New lava dome forming

Assumption 3: Radiometric dating can date the age of rocks very accurately. Radiometric dating is assumed to be practically error-proof by many scientists.

Observation: A few years after Mount St. Helens erupted, a new lava dome formed and hardened. The hardened dome was dated by the potassium-argon radiometric dating method. One of the calculated dates of the dome, which was obviously wrong, was for 350,000 years.

Scientists make assumptions all the time. Assumptions are often correct, but sometimes they are not. The preceding three examples show that it would have been easy to make three wrong assumptions if geologists had examined Mount St. Helens but didn't already know what happened.

When it comes to reconstructing the geological features of the earth, we should give serious consideration to the possibility that catastrophic events associated with the Genesis Flood rapidly changed the face of the earth. These events probably occurred over a few hundred years rather than over millions of years. It's interesting to note that Charles Darwin wrote his famous book *On the Origin of Species* using many of Lyell's ideas about long ages.

MOUNT ST. HELENS

NOTE

Students will probably need some help in understanding that scientists often make assumptions they cannot prove. There are logical reasons for the assumptions scientists make, but they should be re-evaluated as new evidence is introduced. The observable events of Mount St. Helens should cause anyone to consider the possibilities that rock strata and gully systems may have formed quickly and that radiometric dating methods are not always accurate. Remind students that this book is part of a series known as "Investigating the Possibilities." Ask them what it means to investigate the possibilities.

The following additional earth sciences resources are all available through Master Books.

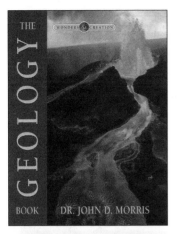

The Geology Book
by Dr. John Morris
978-0-89051-281-4
Grades 6 - 9

Study of earth geology, past and present, from fossilization to volcanic activity.

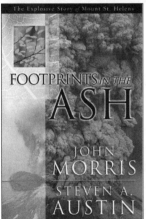

Footprints in the Ash
by Dr. John Morris and
Dr. Steven A. Austin
978-0-89051-400-9
Grades 9 - 12

A comprehensive examination of Mount St. Helens and the testiment it left behind of God's power as evidenced by the Flood of Noah.

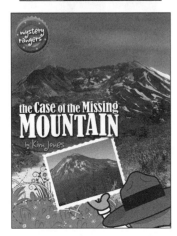

The Case of the Missing Mountain
by Kim Jones
978-0-89051-593-8
Grades 3 - 6

An action-packed mystery that provides explosive fun and stimulating activities surrounding Mount St. Helens.

ADDITIONAL STUDIES: THE GRAND CANYON

Some of the students in Mr. McGregor's science class were giving reports on where they had been during the previous summer. Alice brought some pictures and a book about the Grand Canyon. She said the layers of rocks had taken millions of years to form and they had formed as sediment gradually settled out of ancient oceans that had alternately dried up and reappeared several times. Then it had taken millions of more years for the Colorado River to erode the Grand Canyon down to the oldest kind of rocks on earth. She pointed to a layer of sandstone and said these rocks used to be part of a large desert.

Joey wanted to know how Alice was so sure about how old these places were.

"It says so right here," she replied, holding up a book about the area. "They wouldn't put it in here if it wasn't true."

"How do they know the sandstone rocks used to be a desert?" Joey asked.

"Because it says so right here," Alice repeated firmly.

Joey couldn't wait to talk with his father after school. Dr. Houston taught chemistry at the University, but he knew a lot about geology and methods of dating rocks.

"Hey, Dad, did there used to be a desert in the middle of the Grand Canyon and how old is the Grand Canyon?" Joey got right to the point before he told his dad about Alice and her book.

"Most geologists believe it took millions of years for the Grand Canyon to form, but trying to reconstruct what happened in the past can be tricky."

"Why is that?" Joey asked.

"OK, pay attention or you'll get lost in some big words and the math," his dad smiled.

Find the rest of the story and discussion topics at: www.investigatethepossibilities.org

Mountains *(Folding and Faulting)*

Think about This
Have you ever seen folds in mountains like the ones in the picture?

What do you think might have caused this to happen? Do you think the rocks were hard when they were folded?

The Investigative Problems
What causes rocks to be folded?
What causes rocks to form in flat, level layers?

32

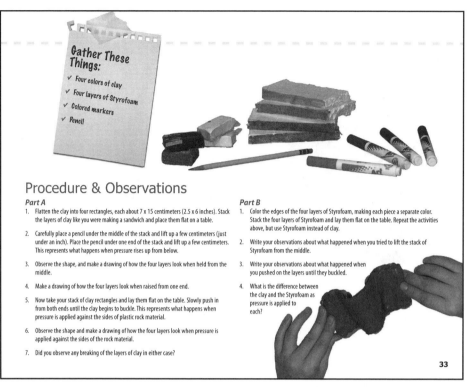

Gather These Things:
✓ Four colors of clay
✓ Four layers of Styrofoam
✓ Colored markers
✓ Pencil

Procedure & Observations

Part A

1. Flatten the clay into four rectangles, each about 7 x 15 centimeters (2.5 x 6 inches). Stack the layers of clay like you were making a sandwich and place them flat on a table.

2. Carefully place a pencil under the middle of the stack and lift up a few centimeters (just under an inch). Place the pencil under one end of the stack and lift up a few centimeters. This represents what happens when pressure rises up from below.

3. Observe the shape, and make a drawing of how the four layers look when held from the middle.

4. Make a drawing of how the four layers look when raised from one end.

5. Now take your stack of clay rectangles and lay them flat on the table. Slowly push in from both ends until the clay begins to buckle. This represents what happens when pressure is applied against the sides of plastic rock material.

6. Observe the shape and make a drawing of how the four layers look when pressure is applied against the sides of the rock material.

7. Did you observe any breaking of the layers of clay in either case?

Part B

1. Color the edges of the four layers of Styrofoam, making each piece a separate color. Stack the four layers of Styrofoam and lay them flat on the table. Repeat the activities above, but use Styrofoam instead of clay.

2. Write your observations about what happened when you tried to lift the stack of Styrofoam from the middle.

3. Write your observations about what happened when you pushed on the layers until they buckled.

4. What is the difference between the clay and the Styrofoam as pressure is applied to each?

33

OBJECTIVES Students should observe examples of where several sedimentary layers are no longer flat and horizontal but have been changed by processes such as folding, tilting, uplifting, or faulting. Other processes that change mountains will be examined.

NOTE Try to help students find and interpret several pictures of layered strata in mountains. Some of the layers will be horizontal, some will be tilted, and some will be bent and folded. Let students give their own explanations for what might have happened. After the lesson is over, have them compare their explanations with the ones given in the book.

The Science Stuff

A sedimentary layer that is laid down out of water in a flat layer is called a "bed." Sedimentary layers can also be laid down by glaciers, wind, or volcanic eruptions, but flat layers are not usually produced in these ways. Most flat layers, such as the ones seen in the Grand Canyon, are laid down by water.

The particles in these layers may become cemented together by certain chemicals as the layers harden. Sedimentary layers that have not completely hardened will often bend without breaking and form folds. The clay models represent this kind of sedimentary layer.

If the layers have hardened and been cemented, they are more likely to break if they are bent, just as you observed that the pieces of Styrofoam broke when they were bent.

Breaks and cracks are commonly found in large rocks and rock formations. When rock on one side of a broken line has slipped and moved, the break is referred to as a fault.

The picture at the beginning of this lesson showed considerable bending, but the rocks aren't broken into pieces or cracked at every bend. A possible explanation is that the layers had not undergone cementation when they were bent. This fits well with the Flood theory that says large amounts of sediment were laid down during a worldwide flood. If the forces pushing on the sedimentary layers had occurred when the sediment had fully hardened, there would be many more cracks in the folded rock layers than there are. However, the layers could have folded easily if they had not yet become cemented. Changes in the rocks from heat and pressure were probably also involved since many of the rocks are metamorphic. Folded rocks still remain part of the mystery.

The three models you made from clay can be observed in nature. There are long stretches of flat layers in the Grand Canyon and the Badlands of South Dakota. Domed layers are seen in the Black Hills. There is an example of an uplifted area on the eastern side of the Grand Canyon. Folded layers are seen in the Rocky Mountains.

Models of what you made with clay

Grand Canyon — long stretches of flat layers

Badlands — long stretches of flat layers

Black Hills — domed layers

Rocky Mountains — folded layers

Making Connections

Roads are sometimes built through mountainous or hilly country. Road crews try to get vegetation to grow on the places that have been cut back to build the road to reduce erosion and weathering of the rocks. When you can find road cuts without vegetation, you may be able to see stratified layers of rocks. Sedimentary layers are very common all around the world. The layers may be flat and horizontal or they may have been changed by processes such as folding, tilting, or uplifting.

34

Dig Deeper

- Make and use a clinometer. This is a useful tool used by geologists as they work in the field, because it allows them to know if the layers are level or if they are at an angle. Take a piece of cardboard slightly larger than the protractor and glue the protractor to it. The straight edge should be on the top and the arch side should be down. Renumber the protractor, starting with 90° on the upper right corner. Go down in 10° increments to 0° at the middle of the bottom of the arch; then go back up the other side in 10° increments till you get to 90° at the upper left corner. Use the pin to attach the pointer. It must swing freely as the clinometer is tilted back and forth. Measure some angles. Then try it out on some exposed rocks and record your measurements.

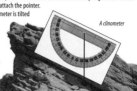

A clinometer

- Find additional examples of rock strata that are found as flat layers, domed layers, uplifted slopes, and folded layers.

What Did You Learn?

1. Give an example of a place in the United States where miles of flat, level layers of strata can be seen.

2. Give an example in the United States where extensive folding of sedimentary layers can be seen.

3. Name at least four ways in which sedimentary layers can be laid down in nature.

4. Which of the following processes is most likely to produce flat level layers of sediment — glaciers, wind, water, or volcanic eruptions?

5. What are some things that can change flat horizontal sedimentary layers in nature after they have been laid down?

6. What do we call breaks and cracks in large rock formations when rock on one side of the crack has slipped and moved?

Pause and Think

Many of the things we read about the Rocky Mountains say that the mountains are millions and millions of years old and contain the fossil remains of many dead plants and animals that died millions and millions of year ago.

Based on the teachings of Genesis, we may find another explanation. We know there was a judgment placed on the earth that was demonstrated by a worldwide Flood. The beginning of this Flood started suddenly with the fountains of the deep bursting forth. Genesis 7:11 says, "In the six hundredth year of Noah's life, on the seventeenth day of the second month — on that day all the springs of the great deep burst forth, and the floodgates of the heavens were opened." This great event would have caused the plates to move quickly and with lots of energy. As the plates moved forcefully against each other, the mountains may have formed. This gives a logical explanation for why the mountains could be young. Dr. Henry M. Morris, who is sometimes known as the father of creation science, believes the Flood is one of the major causes for many of the geological features we can see on the earth.

35

WHAT DID YOU LEARN?

1. Give an example of a place in the United States where miles of flat, level layers of strata can be seen. *Grand Canyon, Badlands of South Dakokta, other places*

2. Give an example in the United States where extensive folding of sedimentary layers can be seen. *Rocky Mountains and other places*

3. Name at least four ways in which sedimentary layers can be laid down in nature. *By water, wind, glaciers, and volcanic eruptions*

4. Which of the following processes is most likely to produce flat level layers of sediment — glaciers, wind, water, or volcanic eruptions? *Water*

5. What are some things that can change flat, horizontal sedimentary layers in nature after they have been laid down? *They may be changed by processes such as folding, tilting, or uplifting.*

6. What do we call breaks and cracks in large rock formations when rock on one side of the crack has slipped and moved? *Faults*

Pardon the Intrusion

Think about This
We've looked at volcanoes that explode and shoot out tons of volcanic ash and other hot materials into the air. We've seen volcanoes that release rivers of hot lava that run down a mountain. But we are limited in what we can observe about what goes on below the ground. We know that magma rises from the mantle through cracks in the earth's crust. Have you ever wondered what happens if hot magma squeezes up into cracks in the crust, but doesn't reach the surface? This actually does happen quite often, but it isn't observable at the time it happens. How do we know this? What's going on below the surface when magma starts to rise?

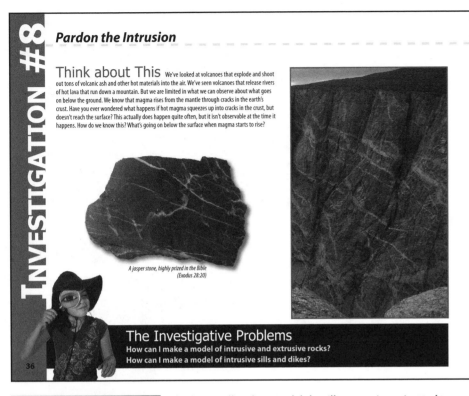

A jasper stone, highly prized in the Bible (Exodus 28:20)

The Investigative Problems
How can I make a model of intrusive and extrusive rocks?
How can I make a model of intrusive sills and dikes?

36

Gather These Things:
✓ Three or four slices of loaf bread with crust trimmed off (or layers of cake)
✓ Peanut butter and jelly mixed together (or dark icing)
✓ Butter knife
✓ Knife to cut the iced layers
✓ Colored toothpicks
✓ Rolled wafers with a dark filling (or narrow flat wafers)

Procedure & Observations

Caution: This activity should be adjusted if any students are diabetic or have peanut allergies. Dispose of all used items when experiment is concluded. Adult supervision required.

Part A
1. Start with three or four layers of soft bread with the crust trimmed off. Spread the peanut mixture between the layers and arrange the slices of bread in a stack like a big sandwich. Use the handle of a clean butter knife to make about four or five vertical holes in the bread that go through all the layers.

2. Make the wafers about a centimeter (just under half an inch) shorter than the height of the iced layers. Gently push a wafer into each of the holes and cover the top with more peanut butter mixture.

3. Cut out a slice of the layers, being sure to cut through the center of one of the wafers. Slice several pieces of the iced layers and observe, and compare the bread model slices to features found in some geological structures.

4. The peanut butter mixture on top represents hardened magma that has reached the surface of the land. The mixture that was not visible before you cut the layers of bread represents hardened magma below the surface. The wafers also represent hardened magma below the surface. The bread represents layers of rock strata.

5. Stick toothpicks in the parts that represent each: red for hardened magma on the surface of the ground; green for hardened magma below the surface of the ground that formed between rock strata; yellow for hardened magma below the surface of the ground that formed through the strata in a (more or less) vertical position; blue for each original strata of rock.

Part B
1. Now read The Science Stuff and identify extrusive rocks, intrusive sills, and intrusive dikes in the bread layers model.

2. What were the red toothpicks?

3. What were the green toothpicks?

4. What were the yellow toothpicks?

5. What were the blue toothpicks?

6. Which rocks were the oldest?

7. Draw two or more of your cutout slices. Label each part where you placed a toothpick.

8. Copy your labeled drawings in your Student Journal.

37

OBJECTIVES Students will make a model that illustrates intrusive and extrusive rocks, as well as sills and dikes.

NOTE Some students may be diabetic or have food allergies. Be especially aware of any students who have peanut allergies or diabetic conditions and avoid any kind of food that might cause problems for students.

As an alternative, using layers of yellow cake and a chocolate icing might be a fun way to do a science lesson. Again, be sure there are no food allergies or diabetic conditions before serving food to students.

The Science Stuff

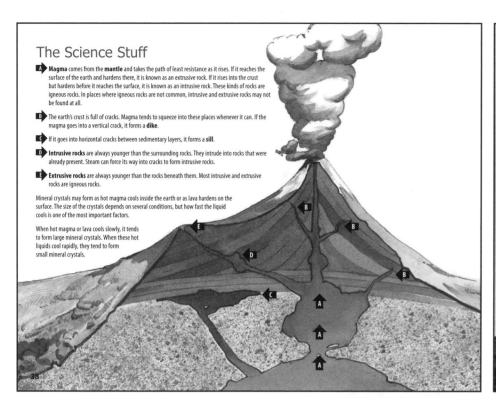

A **Magma** comes from the **mantle** and takes the path of least resistance as it rises. If it reaches the surface of the earth and hardens there, it is known as an extrusive rock. If it rises into the crust but hardens before it reaches the surface, it is known as an intrusive rock. These kinds of rocks are igneous rocks. In places where igneous rocks are not common, intrusive and extrusive rocks may not be found at all.

B The earth's crust is full of cracks. Magma tends to squeeze into these places whenever it can. If the magma goes into a vertical crack, it forms a **dike**.

C If it goes into horizontal cracks between sedimentary layers, it forms a **sill**.

D **Intrusive rocks** are always younger than the surrounding rocks. They intrude into rocks that were already present. Steam can force its way into cracks to form intrusive rocks.

E **Extrusive rocks** are always younger than the rocks beneath them. Most intrusive and extrusive rocks are igneous rocks.

Mineral crystals may form as hot magma cools inside the earth or as lava hardens on the surface. The size of the crystals depends on several conditions, but how fast the liquid cools is one of the most important factors.

When hot magma or lava cools slowly, it tends to form large mineral crystals. When these hot liquids cool rapidly, they tend to form small mineral crystals.

Making Connections

Another unseen feature found below the earth's surface is a magma chamber. These structures contain magma, dissolved gases, and water vapor that have collected in large underground pockets beneath volcanoes. A long vertical crack in the crust, known as a pipe, connects the magma chamber to a vent at the top of a volcano. Other side vents may also branch out from the main pipe. There are many variations of dikes and sills around a volcano.

Dig Deeper

- A number of other features are formed by magma. Moving water, wind, or ice may erode the ground above intrusive rocks so that they can eventually be seen. Choose one or more of the following features and do additional research on them: volcanic necks, batholiths, dome mountains, dikes, and sills. Be sure to include pictures to show as examples.

- Do additional research on a formation in Wyoming known as Devil's Tower. Find more than one explanation for how it formed. Tell which explanation you think best fits the facts.

Detail of a cooled magma chamber in Iceland

Devil's Tower, Wyoming

What Did You Learn?

1. How does magma travel from the mantle through the crust of the earth?
2. When magma reaches the surface of the earth and hardens there, what kind of rock forms?
3. What are intrusive rocks?
4. Are dikes and sills intrusive or extrusive rocks?
5. Suppose you find an intrusive rock that has passed through several layers of stratified rocks. Which would be older — the intrusive rock or the stratified layers of rocks?
6. How does the cooling rate affect the size of mineral crystals that form in intrusive magma?

WHAT DID YOU LEARN?

1. How does magma travel from the mantle through the crust of the earth? *Through cracks in the crust*

2. When magma reaches the surface of the earth and hardens there, what kind of rock forms? *Extrusive rocks*

3. What are intrusive rocks? *Rocks that cool and harden from hot magma below the surface of the earth*

4. Are dikes and sills intrusive or extrusive rocks? *Intrusive*

5. Suppose you find an intrusive rock that has passed through several layers of stratified rocks. Which would be older — the intrusive rock or the stratified layers of rocks? *The stratified layers of rocks*

6. How does the cooling rate affect the size of mineral crystals that form in intrusive magma? *Larger crystals are formed when the magma cools slowly. The crystals are much smaller when the magma cools rapidly.*

Mapping a Mountain

Think about This Olivia's dad, who was a surveyor, was showing the family the map he had produced of the area where they lived. He pointed to their house and to their neighbor's house. Olivia wanted to know what all the curved lines were on the map.

"Those are contour lines," he explained. "You can see from the map that our house is built on the highest part of one of the hills. Your friend Kelly lives on another hill that is 20 feet higher than ours."

How do you think Olivia's dad knew this just from looking at a map?

Why don't you make a "mountain" out of clay and you can find out!

Procedure & Observations

Part A

1. Mold the clay into the shape of a mountain that is 8 to 10 centimeters (3 to 4 inches) high. Make a steep cliff at the top of one side. Make a gradual slope somewhere on one side. Try to make it look like a real mountain would look. Refer to a picture of a mountain if you need to.

2. Set the "mountain" on a piece of paper. Put the paper and "mountain" on a table and draw a line with your marker around the bottom edge of your "mountain." Now place a block next to the "mountain."

3. Lay a flat ruler on the top of the block and point one end toward the "mountain." Holding the ruler firmly on the block, put a mark on the side of the "mountain" and slide the block around the "mountain."

4. Go all the way around it, making a line in the clay at the same level. Reinforce the line with a marker. Stand over your "mountain" and look down on the line you just drew. This is a contour line.

The Investigative Problems
What is a topographic map?
How do contour lines show elevation?

40

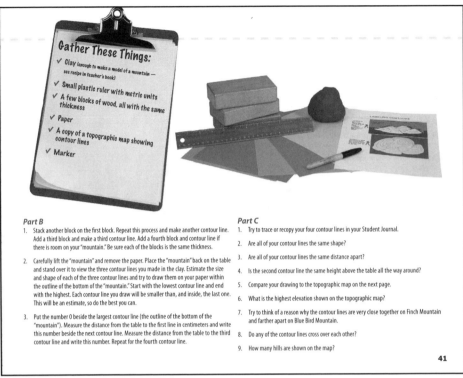

Gather These Things:

✓ Clay (enough to make a model of a mountain — see recipe in teacher's book)
✓ Small plastic ruler with metric units
✓ A few blocks of wood, all with the same thickness
✓ Paper
✓ A copy of a topographic map showing contour lines
✓ Marker

Part B

1. Stack another block on the first block. Repeat this process and make another contour line. Add a third block and make a third contour line. Add a fourth block and contour line if there is room on your "mountain." Be sure each of the blocks is the same thickness.

2. Carefully lift the "mountain" and remove the paper. Place the "mountain" back on the table and stand over it to view the three contour lines you made in the clay. Estimate the size and shape of each of the three contour lines and try to draw them on your paper within the outline of the bottom of the "mountain." Start with the lowest contour line and end with the highest. Each contour line you draw will be smaller than, and inside, the last one. This will be an estimate, so do the best you can.

3. Put the number 0 beside the largest contour line (the outline of the bottom of the "mountain"). Measure the distance from the table to the first line in centimeters and write this number beside the next contour line. Measure the distance from the table to the third contour line and write this number. Repeat for the fourth contour line.

Part C

1. Try to trace or recopy your four contour lines in your Student Journal.

2. Are all of your contour lines the same shape?

3. Are all of your contour lines the same distance apart?

4. Is the second contour line the same height above the table all the way around?

5. Compare your drawing to the topographic map on the next page.

6. What is the highest elevation shown on the topographic map?

7. Try to think of a reason why the contour lines are very close together on Finch Mountain and farther apart on Blue Bird Mountain.

8. Do any of the contour lines cross over each other?

9. How many hills are shown on the map?

41

OBJECTIVES

Students will make their own contour maps of a model of a mountain.

They should learn how to read contour maps and understand the symbols on them.

NOTE

You can make your own clay from flour, salt, water, and oil. There are several recipes on the Internet, but this is one that is simple. Thoroughly mix together two parts flour, one part salt, one part water, and a little oil.

The contour maps drawn by the students may not be exact. Some guessing will be necessary. The main thing is for students to gain an understanding of contour maps. The maps should show four contour lines, each one drawn smaller than the one before.

Some topographic maps include contour lines plus other detailed information. You may want to help students learn some of these symbols, but don't be concerned about too many details.

As students examine the map (procedure), be sure they understand the map shows two mountains. Finch Mountain is about 600 feet high, and Blue Bird Mountain is about 400 feet high. If 300 feet is written by a contour line, that means the elevation is 300 feet everywhere on that line.

The Science Stuff

Geologists have used contour lines for many years as a means of showing elevation of landforms on a flat map. Each contour line makes a complete pathway and never crosses another contour line. Every place on that line is at the same elevation. The contour line next to it will represent a different elevation that is a little more or a little less.

When the contour lines are very close together, the map is showing a steep slope. When the lines are farther apart, the map is showing a more gradual slope.

Topographic maps generally show contour lines, as well as other details of an area. There may be symbols for highways, railroads, streams, lakes, wooded areas, houses, and other features on a topographic map.

The numbers on the contour lines usually show the elevation relative to sea level. The elevation number will be the same all the way around a contour line. The increase or decrease in height between any two contour lines on a map is the same. Examine the topographic map below to see how they are created.

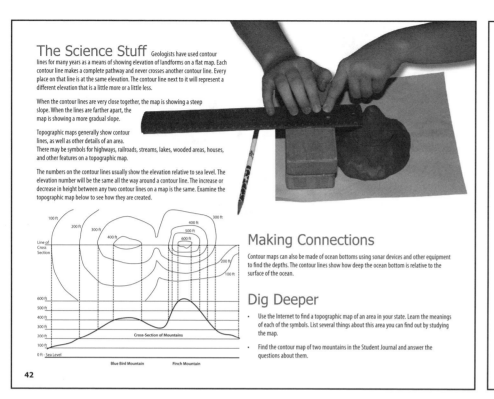

Making Connections

Contour maps can also be made of ocean bottoms using sonar devices and other equipment to find the depths. The contour lines show how deep the ocean bottom is relative to the surface of the ocean.

Dig Deeper

- Use the Internet to find a topographic map of an area in your state. Learn the meanings of each of the symbols. List several things about this area you can find out by studying the map.

- Find the contour map of two mountains in the Student Journal and answer the questions about them.

42

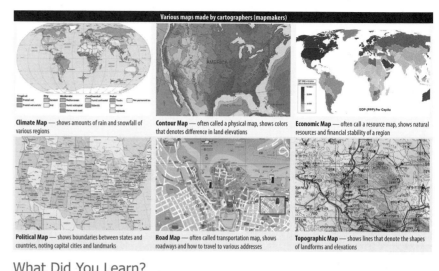

Various maps made by cartographers (mapmakers)

Climate Map — shows amounts of rain and snowfall of various regions

Contour Map — often called a physical map, shows colors that denotes difference in land elevations

Economic Map — often call a resource map, shows natural resources and financial stability of a region

Political Map — shows boundaries between states and countries, noting capital cities and landmarks

Road Map — often called transportation map, shows roadways and how to travel to various addresses

Topographic Map — shows lines that denote the shapes of landforms and elevations

What Did You Learn?

1. If the number 990 feet is written next to a certain contour line on a map, what does that mean?

2. Do contour lines overlap each other?

3. What do contour lines on a map that are close together indicate? What do contour lines that are spaced farther apart indicate?

4. What kind of map shows contour lines, as well as other details of specific features in an area?

5. Is the increase or decrease in height between any two contour lines on a map the same?

6. Whether a contour map shows the depth of the ocean bottom or the elevation of dry land, what is the starting point for measuring these distances?

43

WHAT DID YOU LEARN?

1. If the number 990 feet is written next to a certain contour line on a map, what does that mean? *It means that every place along that line is 990 feet above sea level.*

2. Do contour lines overlap each other? *No*

3. What do contour lines on a map that are close together indicate? *A steep incline or hill* What do contour lines that are spaced farther apart indicate? *A gradual incline or hill*

4. What kind of map shows contour lines, as well as other details of specific features in an area? *Topographic map*

5. Is the increase or decrease in height between any two contour lines on a map the same? *Yes*

6. Whether a contour map shows the depth of the ocean bottom or the elevation of dry land, what is the starting point for measuring these distances? *The surface of the ocean (sea level)*

INVESTIGATION #10

Growing Crystals

Think about This
The sign in the store window simply said ROCKS AND MINERALS. Rachel's aunt invited her to look at the pink rocks in the window. They had flat sides and came to a point at the top of each. Her aunt explained that they weren't really rocks. They were actually a type of mineral known as quartz, but they were collected in the mountains and they were produced naturally. She asked Rachel if she would like to grow her own crystals at home. What kind of crystals do you think Rachel would be able to grow at home? Would they be as pretty as the pink quartz crystals in the store window?

The Investigative Problems
How can you make crystals?
Do different kinds of chemicals produce crystals with different shapes?

44

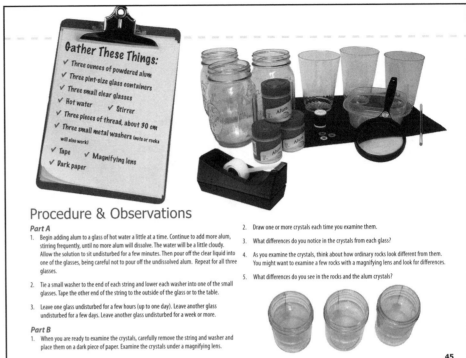

Gather These Things:
- ✓ Three ounces of powdered alum
- ✓ Three pint-size glass containers
- ✓ Three small clear glasses
- ✓ Hot water   ✓ Stirrer
- ✓ Three pieces of thread, about 30 cm
- ✓ Three small metal washers (nuts or rocks will also work)
- ✓ Tape   ✓ Magnifying lens
- ✓ Dark paper

Procedure & Observations

Part A

1. Begin adding alum to a glass of hot water a little at a time. Continue to add more alum, stirring frequently, until no more alum will dissolve. The water will be a little cloudy. Allow the solution to sit undisturbed for a few minutes. Then pour off the clear liquid into one of the glasses, being careful not to pour off the undissolved alum. Repeat for all three glasses.

2. Tie a small washer to the end of each string and lower each washer into one of the small glasses. Tape the other end of the string to the outside of the glass or to the table.

3. Leave one glass undisturbed for a few hours (up to one day). Leave another glass undisturbed for a few days. Leave another glass undisturbed for a week or more.

Part B

1. When you are ready to examine the crystals, carefully remove the string and washer and place them on a dark piece of paper. Examine the crystals under a magnifying lens.

2. Draw one or more crystals each time you examine them.

3. What differences do you notice in the crystals from each glass?

4. As you examine the crystals, think about how ordinary rocks look different from them. You might want to examine a few rocks with a magnifying lens and look for differences.

5. What differences do you see in the rocks and the alum crystals?

45

OBJECTIVES

Students will grow their own crystals and observe their characteristics.

They will understand the conditions necessary to grow crystals.

NOTE

Students usually enjoy watching crystals grow. Be sure to find a place where the crystals can sit undisturbed for a few days. If students show an interest in trying to grow other kinds of crystals, they can find a variety of instructions on the Internet or other sources. It's a good idea to avoid the ones that are poisonous unless carefully supervised by an adult.

The Science Stuff

Crystals are pure substances. Each kind of crystal is composed of one kind of chemical that is arranged in a particular repeating pattern.

As the saturated solution of alum cools and evaporates, alum crystals may begin to form. A saturated solution is one that has dissolved as much of a chemical as possible at a particular temperature. Instead of settling out of the solution, the alum has time to slowly organize itself as a crystal. It will begin growing around a "seed."

You probably found alum crystals on the string and maybe in other places in the glass as well. They will start out small and get larger if the right conditions are present.

Except for a few minerals, each kind of crystal has a distinct geometric shape. Crystals have flat, smooth sides called faces, while rocks have irregular shapes. This is one of the main ways in which crystals are different from rocks. Alum crystals have a diamond shape, regardless of their size. Salt crystals are cube-shaped.

Crystals also form in nature under similar conditions as those used to grow alum crystals. Crystals and crystalline structures that are found in nature are known as minerals. They often begin in hot, saturated solutions. Crystals form as the surrounding liquid cools and/or evaporates.

You probably noticed that the alum crystals that were left undisturbed for a week were larger than the ones that were left for a few hours. Generally, the size of crystals depends on the time they are allowed to grow. Small crystals tend to form when the liquid cools rapidly, and larger crystals tend to form when the liquid cools slowly.

Different geometrically shaped crystals

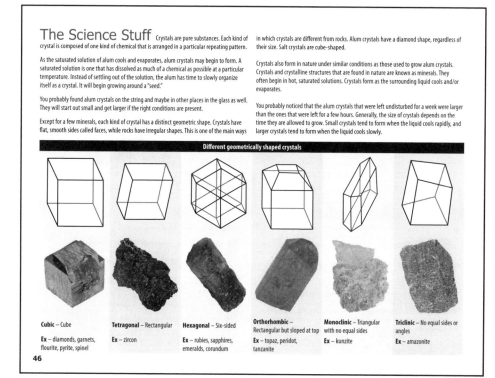

Cubic – Cube
Ex – diamonds, garnets, flourite, pyrite, spinel

Tetragonal – Rectangular
Ex – zircon

Hexagonal – Six-sided
Ex – rubies, sapphires, emeralds, corundum

Orthorhombic – Rectangular but sloped at top
Ex – topaz, peridot, tanzanite

Monoclinic – Triangular with no equal sides
Ex – kunzite

Triclinic – No equal sides or angles
Ex – amazonite

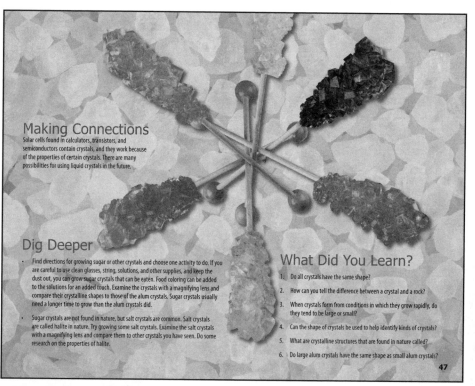

Making Connections

Solar cells found in calculators, transistors, and semiconductors contain crystals, and they work because of the properties of certain crystals. There are many possibilities for using liquid crystals in the future.

Dig Deeper

- Find directions for growing sugar or other crystals and choose one activity to do. If you are careful to use clean glasses, string, solutions, and other supplies, and keep the dust out, you can grow sugar crystals that can be eaten. Food coloring can be added to the solutions for an added touch. Examine the crystals with a magnifying lens and compare their crystalline shapes to those of the alum crystals. Sugar crystals usually need a longer time to grow than the alum crystals did.

- Sugar crystals are not found in nature, but salt crystals are common. Salt crystals are called halite in nature. Try growing some salt crystals. Examine the salt crystals with a magnifying lens and compare them to other crystals you have seen. Do some research on the properties of halite.

What Did You Learn?

1. Do all crystals have the same shape?
2. How can you tell the difference between a crystal and a rock?
3. When crystals form from conditions in which they grow rapidly, do they tend to be large or small?
4. Can the shape of crystals be used to help identify kinds of crystals?
5. What are crystalline structures that are found in nature called?
6. Do large alum crystals have the same shape as small alum crystals?

WHAT DID YOU LEARN?

1. Do all crystals have the same shape? *No*

2. How can you tell the difference in a crystal and a rock? *Crystals have flat faces and a definite shape and rocks don't. Crystals are pure substances and rocks are a mixture of minerals (crystals).*

3. When crystals form from conditions in which they grow rapidly, do they tend to be large or small? *Small*

4. Can the shape of crystals be used to help identify kinds of crystals? *Yes*

5. What are crystalline structures that are found in nature called? *Minerals*

6. Do large alum crystals have the same shape as small alum crystals? *Yes (same shape, but not necessarily the same size)*

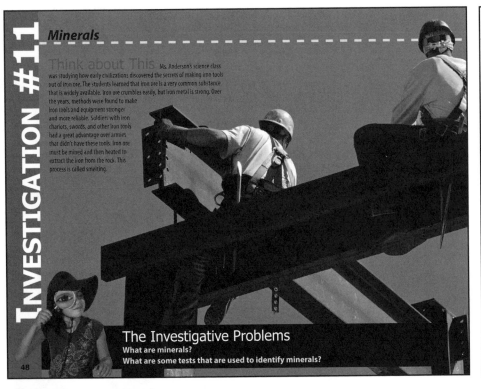

INVESTIGATION #11

Minerals

Think about This

Ms. Anderson's science class was studying how early civilizations discovered the secrets of making iron tools out of iron ore. The students learned that iron ore is a very common substance that is widely available. Iron ore crumbles easily, but iron metal is strong. Over the years, methods were found to make iron tools and equipment stronger and more reliable. Soldiers with iron chariots, swords, and other iron tools had a great advantage over armies that didn't have these tools. Iron ore must be mined and then heated to extract the iron from the rock. This process is called smelting.

The Investigative Problems

What are minerals?
What are some tests that are used to identify minerals?

48

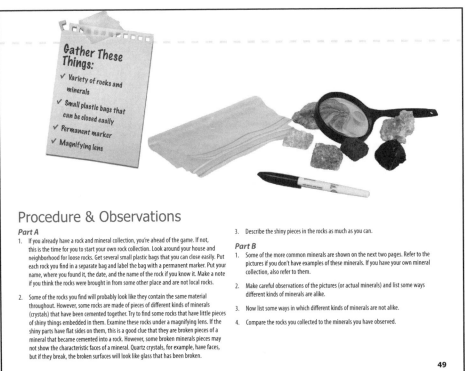

Gather These Things:

✓ Variety of rocks and minerals
✓ Small plastic bags that can be closed easily
✓ Permanent marker
✓ Magnifying lens

Procedure & Observations

Part A

1. If you already have a rock and mineral collection, you're ahead of the game. If not, this is the time for you to start your own rock collection. Look around your house and neighborhood for loose rocks. Get several small plastic bags that you can close easily. Put each rock you find in a separate bag and label the bag with a permanent marker. Put your name, where you found it, the date, and the name of the rock if you know it. Make a note if you think the rocks were brought in from some other place and are not local rocks.

2. Some of the rocks you find will probably look like they contain the same material throughout. However, some rocks are made of pieces of different kinds of minerals (crystals) that have been cemented together. Try to find some rocks that have little pieces of shiny things embedded in them. Examine these rocks under a magnifying lens. If the shiny parts have flat sides on them, this is a good clue that they are broken pieces of a mineral that became cemented into a rock. However, some broken minerals pieces may not show the characteristic faces of a mineral. Quartz crystals, for example, have faces, but if they break, the broken surfaces will look like glass that has been broken.

3. Describe the shiny pieces in the rocks as much as you can.

Part B

1. Some of the more common minerals are shown on the next two pages. Refer to the pictures if you don't have examples of these minerals. If you have your own mineral collection, also refer to them.

2. Make careful observations of the pictures (or actual minerals) and list some ways different kinds of minerals are alike.

3. Now list some ways in which different kinds of minerals are not alike.

4. Compare the rocks you collected to the minerals you have observed.

49

OBJECTIVES

Students will become familiar with some of the common tests that are used to identify rocks and minerals.

NOTE

It will be helpful to have a mineral and rock collection for the next few lessons to compare rocks found by students with rocks and minerals that have been correctly identified. There are also good Internet sites and book references that show and write about the chemical and physical properties of many rocks and minerals. Encourage students to begin a collection of rocks and minerals even if they can't immediately identify every one. Chances are good that they will be able to identify several or all within the next few months.

The Science Stuff

When crystals are found in nature, they are called minerals. A substance is classified as a mineral if it forms in nature, doesn't contain remains of living things, and has a definite crystal structure and chemical make-up.

Small broken pieces of minerals are found in many rocks. Granite is a rock that contains pieces of quartz, mica, and feldspar that have been cemented together. The percentages of these minerals may vary quite a bit in different sources of granite.

Minerals can be identified by performing a series of tests to determine certain physical and chemical properties. Usually more than one test will need to be performed in order to absolutely identify a specific mineral. The most common tests are:

1. color test
2. streak test
3. luster
4. crystal form
5. cleavage test
6. hardness test (Mohs's scale of hardness)
7. density test

There are other tests that can help identify minerals, such as magnetic properties, fluorescent glow under ultraviolet light, radioactivity, formation of bubbles when exposed to weak hydrochloric acid, flame test, etc.

The most common minerals are the silicate minerals, which make up more than 95 percent of the earth's crust. These minerals are made up of compounds of silicon and oxygen. The main minerals that make up rocks are feldspars, micas, olivines, pyroxenes, amphiboles, quartz, clay minerals, and calcite (calcium carbonate).

Here's one explanation of how a mineral may form in nature. The solid part of the earth's mantle becomes liquid magma when the pressure is reduced. This can happen when there is a crack or break in the earth's crust. The liquid magma may work its way upward through openings in the rock above.

The liquid magma may travel part of the way toward the surface and then begin to cool. Sometimes the magma reaches the surface of the earth. If the cooling process is slow, there is an opportunity for large crystals to form. If the cooling process is rapid, glassy rocks or very fine crystals will probably form.

Beautiful quartz crystals have been found in nature that are several inches long. These large crystals probably took much longer to "grow" than the more common smaller ones.

Not all minerals crystallize out of hot molten liquids. Minerals may also crystallize out of saturated solutions of chemicals, much like you made crystals in the previous lesson.

Three varieties of granite

Pyrite

Feldspar

Mica

Gypsum

Quartz

50

What Did You Learn?

1. When crystals are found in nature, what are they called?
2. Why is granite classified as a rock instead of a mineral?
3. Are minerals and crystals pure substances or mixtures of chemicals?
4. Name the minerals found in granite.
5. Name at least eight tests that can be done to identify minerals.
6. What kinds of minerals make up about 95 percent of the minerals in the earth's crust?
7. Name at least eight minerals that make up most of the rocks in the earth's crust.

Making Connections

Gemstones are minerals that are somewhat rare. Diamonds, rubies, and emeralds are gemstones that are greatly valued. When large gemstones are found without impurities and other flaws, they are quite valuable.

Sometimes diamonds have other uses than to be made into jewelry. They are also used in drills and other machinery because of their hardness.

Dig Deeper

- Find the names of at least 10 gemstones. Find a picture and some information about each one.
- Begin a collection of minerals.
- Synthetic diamonds and rubies can be produced. Find out what kinds of things they are used in.

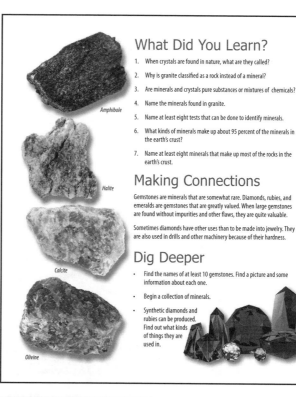

Amphibole

Halite

Calcite

Olivine

Pause and Think

Genesis 1:9–10 indicates that the earth's crust formed on day 3. "And God said, 'Let the waters under the heavens be gathered together into one place, and let dry land appear'; and it was so. And God called the dry land Earth, and the gathering together of the waters He called Seas. And God saw that it was good."

Notice that first the waters were gathered together into one place, and second, dry land appeared. One logical explanation for how the massive crystalline solid base of the crust formed during day 3 is that it came out of these waters. We know that crystals could have formed and separated out of a saturated liquid solution. Also, hot magma rising into the crust could have cooled and formed crystals. Locked-up water vapor may have had an important role in making these crystalline rocks. Even today, magma in active volcanoes releases large amounts of water vapor. No one knows for sure the processes God used, but we do know that day 3 was part of God's good design and plan.

51

WHAT DID YOU LEARN?

1. When crystals are found in nature, what are they called? *Minerals*

2. Why is granite classified as a rock instead of a mineral? *Granite is made up of several minerals that have been cemented together into a rock. A mineral is a pure substance.*

3. Are minerals and crystals pure substances or mixtures of chemicals? *Minerals and crystals are pure substances.*

4. Names the minerals found in granite. *Granite contains pieces of quartz, mica, and feldspar.*

5. Name at least eight tests that can be done to identify minerals. *Color test, streak test, luster, crystal form, cleavage test, hardness test (Mohs's scale of hardness), density test, test for magnetic properties, fluorescent glow under ultraviolet light, radioactivity, formation of bubbles when exposed to weak acid, flame test, and others.*

6. What kinds of minerals make up about 95 percent of the minerals in the earth's crust? *Silicate minerals (composed of silicon and oxygen chemically joined together)*

7. Name at least eight minerals that make up most of the rocks in the earth's crust. *Feldspars, micas, olivines, pyroxenes, amphiboles, quartz, clay minerals, and calcite (calcium carbonate)*

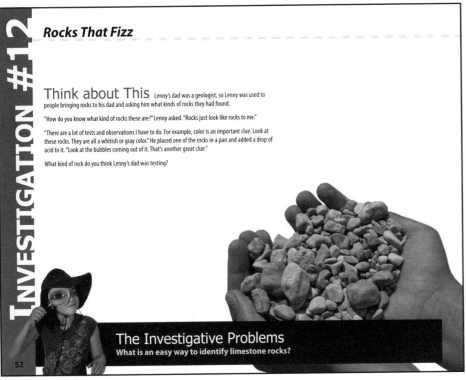

INVESTIGATION #12

Rocks That Fizz

Think about This

Lenny's dad was a geologist, so Lenny was used to people bringing rocks to his dad and asking him what kinds of rocks they had found.

"How do you know what kind of rocks these are?" Lenny asked. "Rocks just look like rocks to me."

"There are a lot of tests and observations I have to do. For example, color is an important clue. Look at these rocks. They are all a whitish or gray color." He placed one of the rocks in a pan and added a drop of acid to it. "Look at the bubbles coming out of it. That's another great clue."

What kind of rock do you think Lenny's dad was testing?

The Investigative Problems
What is an easy way to identify limestone rocks?

52

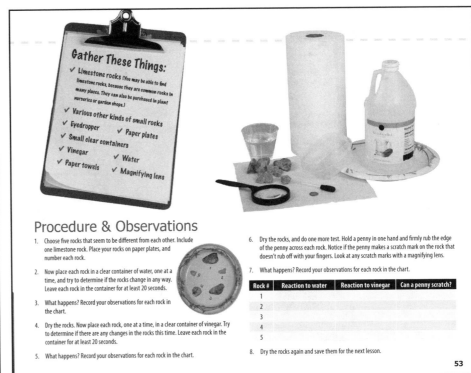

Gather These Things:
- ✓ Limestone rocks (You may be able to find limestone rocks, because they are common rocks in many places. They can also be purchased in plant nurseries or garden shops.)
- ✓ Various other kinds of small rocks
- ✓ Eyedropper   ✓ Paper plates
- ✓ Small clear containers
- ✓ Vinegar   ✓ Water
- ✓ Paper towels   ✓ Magnifying lens

Procedure & Observations

1. Choose five rocks that seem to be different from each other. Include one limestone rock. Place your rocks on paper plates, and number each rock.

2. Now place each rock in a clear container of water, one at a time, and try to determine if the rocks change in any way. Leave each rock in the container for at least 20 seconds.

3. What happens? Record your observations for each rock in the chart.

4. Dry the rocks. Now place each rock, one at a time, in a clear container of vinegar. Try to determine if there are any changes in the rocks this time. Leave each rock in the container for at least 20 seconds.

5. What happens? Record your observations for each rock in the chart.

6. Dry the rocks, and do one more test. Hold a penny in one hand and firmly rub the edge of the penny across each rock. Notice if the penny makes a scratch mark on the rock that doesn't rub off with your fingers. Look at any scratch marks with a magnifying lens.

7. What happens? Record your observations for each rock in the chart.

Rock #	Reaction to water	Reaction to vinegar	Can a penny scratch?
1			
2			
3			
4			
5			

8. Dry the rocks again and save them for the next lesson.

53

OBJECTIVES

Students will do a chemical test on limestone rock and observe that limestone will produce bubbles when an acid is placed on it. Most all other kinds of rocks will not react in this way.

NOTE

If you live near an area where there are light-colored, easily scratched rocks next to seashell fossils, you can be reasonably sure these are limestone rocks. However, fossils may not always be readily seen in limestone. To test, put the rocks in a small container of vinegar. Limestone will quickly begin to form bubbles. If you're still not sure, go to a place that sells garden/nursery supplies and purchase a small container of decorative limestone rocks.

The Science Stuff

Identifying rocks is a matter of doing a series of tests and observations and looking for clues. Color is a clue, but some kinds of rocks come in many colors. Other tests and observations will be needed to correctly identify a rock.

A combination of three clues will usually be enough to identify limestone, although you may need to do a few more tests to be sure.

(1) Limestone is most often a whitish or gray color, although impurities in the rock may give it other colors. (2) You can rub a copper penny forcefully across it and it will leave a scratch mark. (3) You can put a limestone rock in a container of vinegar and it will make bubbles.

If your light-colored rock begins to bubble when you add vinegar, it is probably limestone. If no bubbles are produced, it is probably not limestone.

The bubbles contain carbon dioxide. They come from a chemical reaction between calcium carbonate and vinegar (an acid) in which carbon dioxide gas is formed. The chemical formula for calcium carbonate is $CaCO_3$.

The main mineral in limestone rock is calcite, which is made of calcium carbonate. Magnesium carbonate is widespread in sedimentary rocks and will also produce a few bubbles under the right conditions, but not as fast as calcium carbonate will.

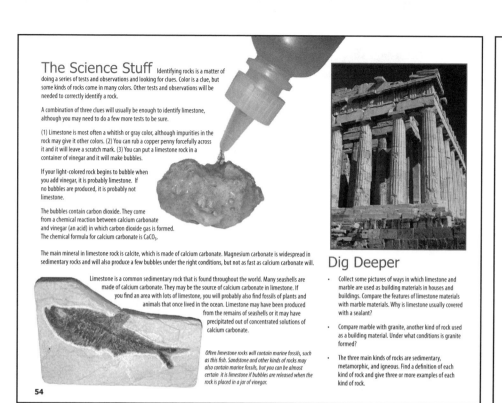

Limestone is a common sedimentary rock that is found throughout the world. Many seashells are made of calcium carbonate. They may be the source of calcium carbonate in limestone. If you find an area with lots of limestone, you will probably also find fossils of plants and animals that once lived in the ocean. Limestone may have been produced from the remains of seashells or it may have precipitated out of concentrated solutions of calcium carbonate.

Often limestone rocks will contain marine fossils, such as this fish. Sandstone and other kinds of rocks may also contain marine fossils, but you can be almost certain it is limestone if bubbles are released when the rock is placed in a jar of vinegar.

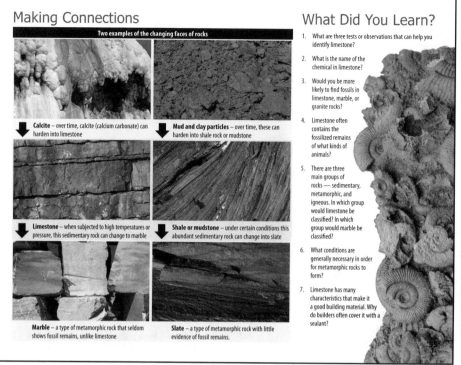

Dig Deeper

- Collect some pictures of ways in which limestone and marble are used as building materials in houses and buildings. Compare the features of limestone materials with marble materials. Why is limestone usually covered with a sealant?

- Compare marble with granite, another kind of rock used as a building material. Under what conditions is granite formed?

- The three main kinds of rocks are sedimentary, metamorphic, and igneous. Find a definition of each kind of rock and give three or more examples of each kind of rock.

54

Making Connections

Two examples of the changing faces of rocks

Calcite – over time, calcite (calcium carbonate) can harden into limestone

Mud and clay particles – over time, these can harden into shale rock or mudstone

Limestone – when subjected to high temperatures or pressure, this sedimentary rock can change to marble

Shale or mudstone – under certain conditions this abundant sedimentary rock can change into slate

Marble – a type of metamorphic rock that seldom shows fossil remains, unlike limestone

Slate – a type of metamorphic rock with little evidence of fossil remains.

What Did You Learn?

1. What are three tests or observations that can help you identify limestone?

2. What is the name of the chemical in limestone?

3. Would you be more likely to find fossils in limestone, marble, or granite rocks?

4. Limestone often contains the fossilized remains of what kinds of animals?

5. There are three main groups of rocks — sedimentary, metamorphic, and igneous. In which group would limestone be classified? In which group would marble be classified?

6. What conditions are generally necessary in order for metamorphic rocks to form?

7. Limestone has many characteristics that make it a good building material. Why do builders often cover it with a sealant?

WHAT DID YOU LEARN?

1. **What are three tests or observations that can help you identify limestone?** *(1) Limestone is most often a whitish or gray color. (2) You can rub a copper penny forcefully across it and it will leave a scratch mark. (3) You can put a limestone rock in a container of vinegar and it will make bubbles.*

2. **What is the name of the chemical in limestone?** *Calcium carbonate*

3. **Would you be more likely to find fossils in limestone, marble, or granite rocks?** *Limestone (a sedimentary rock)*

4. **Limestone often contains the fossilized remains of what kinds of animals?** *Plants and animals that live in the ocean (or once lived there)*

5. **There are three main groups of rocks — sedimentary, metamorphic, and igneous. In which group would limestone be classified? In which group would marble be classified?** *Limestone is classified as a sedimentary rock. Marble is classified as a metamorphic rock.*

6. **What conditions are generally necessary in order for metamorphic rocks to form?** *Heat and pressure*

7. **Limestone has many characteristics that make it a good building material. Why do builders often cover it with a sealant?** *It can be eroded by acids. (Carbonic acid and vinegar are common acids in homes.)*

Rocks Have an ID

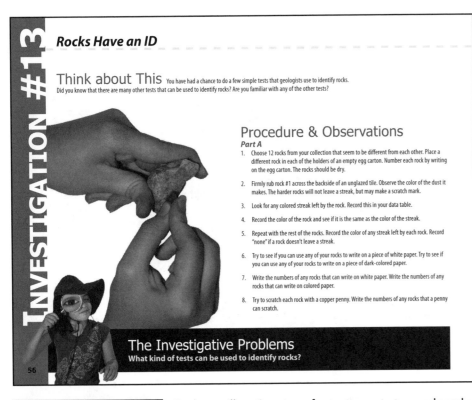

Think about This
You have had a chance to do a few simple tests that geologists use to identify rocks. Did you know that there are many other tests that can be used to identify rocks? Are you familiar with any of the other tests?

Procedure & Observations
Part A
1. Choose 12 rocks from your collection that seem to be different from each other. Place a different rock in each of the holders of an empty egg carton. Number each rock by writing on the egg carton. The rocks should be dry.

2. Firmly rub rock #1 across the backside of an unglazed tile. Observe the color of the dust it makes. The harder rocks will not leave a streak, but may make a scratch mark.

3. Look for any colored streak left by the rock. Record this in your data table.

4. Record the color of the rock and see it is the same as the color of the streak.

5. Repeat with the rest of the rocks. Record the color of any streak left by each rock. Record "none" if a rock doesn't leave a streak.

6. Try to see if you can use any of your rocks to write on a piece of white paper. Try to see if you can use any of your rocks to write on a piece of dark-colored paper.

7. Write the numbers of any rocks that can write on white paper. Write the numbers of any rocks that can write on colored paper.

8. Try to scratch each rock with a copper penny. Write the numbers of any rocks that a penny can scratch.

The Investigative Problems
What kind of tests can be used to identify rocks?

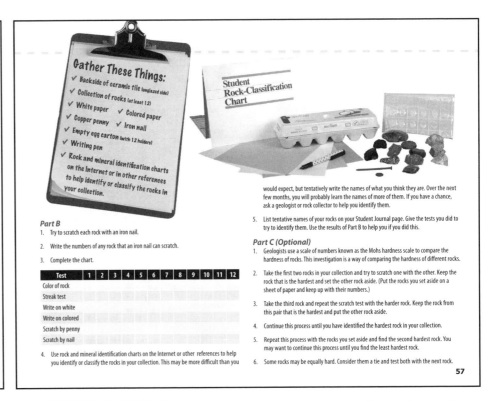

Gather These Things:
- ✓ Backside of ceramic tile (unglazed side)
- ✓ Collection of rocks (at least 12)
- ✓ White paper   ✓ Colored paper
- ✓ Copper penny   ✓ Iron nail
- ✓ Empty egg carton (with 12 holders)
- ✓ Writing pen
- ✓ Rock and mineral identification charts on the Internet or in other references to help identify or classify the rocks in your collection.

would expect, but tentatively write the names of what you think they are. Over the next few months, you will probably learn the names of more of them. If you have a chance, ask a geologist or rock collector to help you identify them.

5. List tentative names of your rocks on your Student Journal page. Give the tests you did to try to identify them. Use the results of Part B to help you if you did this.

Part B
1. Try to scratch each rock with an iron nail.

2. Write the numbers of any rock that an iron nail can scratch.

3. Complete the chart.

Test	1	2	3	4	5	6	7	8	9	10	11	12
Color of rock												
Streak test												
Write on white												
Write on colored												
Scratch by penny												
Scratch by nail												

4. Use rock and mineral identification charts on the Internet or other references to help you identify or classify the rocks in your collection. This may be more difficult than you

Part C (Optional)
1. Geologists use a scale of numbers known as the Mohs hardness scale to compare the hardness of rocks. This investigation is a way of comparing the hardness of different rocks.

2. Take the first two rocks in your collection and try to scratch one with the other. Keep the rock that is the hardest and set the other rock aside. (Put the rocks you set aside on a sheet of paper and keep up with their numbers.)

3. Take the third rock and repeat the scratch test with the harder rock. Keep the rock from this pair that is the hardest and put the other rock aside.

4. Continue this process until you have identified the hardest rock in your collection.

5. Repeat this process with the rocks you set aside and find the second hardest rock. You may want to continue this process until you find the least hardest rock.

6. Some rocks may be equally hard. Consider them a tie and test both with the next rock.

OBJECTIVES
Students will continue to perform common tests on rocks and minerals. Hopefully, they will be able to identify some of the rocks and minerals they have collected.

NOTE
Streak tests are interesting to students. The streak is not always the same color as the rock. A streak test combined with a few other clues will enable students to identify many rocks and minerals.

Rocks and minerals have different definitions, but it is not necessary to make too many distinctions between them at this level. In nature, the terms tend to be used interchangeably unless there are clear-cut crystal formations. Pieces of minerals can often be seen cemented together in rocks.

The Science Stuff

You should recall the list of tests mentioned in Investigation #10 that can be used to identify minerals. The term "rock" may refer loosely to both rocks and minerals. In fact, rocks are made up of minerals and are sometimes a mixture of several minerals. Some of these tests can be used to identify both rocks and minerals.

The streak test is one of the most common tests used in classifying rocks. The color of the steak is frequently different from the color of the rock.

The scratch test can help you by eliminating rocks that are too "soft" or rocks that are too hard. The acid test you did in Investigation #11 can help you identify limestone rocks.

Recall from the previous chapter that rocks can be classified into three big groups: igneous rocks, sedimentary rocks, and metamorphic rocks.

Igneous rocks are formed from the cooling of magma or other melted rock inside the earth. If the rock formed from magma reaches the surface of the earth and then cools and hardens, it is known as an extrusive rock. If the magma rises from the mantle but doesn't reach the earth's surface, it is known as an intrusive rock.

Some types of igneous rock include lava, basalt, granite, and obsidian. Granite is the most common igneous intrusive rock. It is composed of the minerals quartz, feldspar, and mica. The combination of these minerals gives the rock a speckled appearance.

Sedimentary rocks are rocks that are formed from broken pieces of other rocks. Most sedimentary rocks are laid down by water. These kinds of rocks contain billions of fossils and are found all over the world. Examples of sedimentary rocks include sandstone, limestone, shale, and conglomerate.

Metamorphic rocks are formed when igneous rocks or sedimentary rocks are changed by high temperatures and/or pressure into new kinds of rock. Examples of metamorphic rocks made from igneous rocks are gneiss (pronounced "nice") and schist. Metamorphic rocks made from sedimentary rocks include slate made from shale, marble made from limestone, and quartzite made from sandstone.

Dig Deeper

Build a rock collection. Start with rocks in your own area. Look for rocks when you travel to other places. Ask people who go on vacations to bring you rocks. Be sure to keep records of the rocks you find. Small plastic bags labeled with permanent markers are one way. Some people keep rocks in egg cartons that are numbered. The information is kept in a notebook. Even if you don't know the names of the rocks at first, you may be able to find someone who can help you identify them later.

Making Connections

The Mohs hardness scale uses the following minerals to compare the hardness of minerals and rocks — "1" is the least hard mineral and "10" is the hardest mineral. Notice that quartz has a hardness of 7. Quartz can scratch any mineral with a hardness of less than 7, but it cannot scratch a mineral with a hardness greater than 7.

Mohs Hardness Scale

1	talc	6	orthoclase feldspar
2	gypsum	7	quartz
3	calcite	8	topaz
4	fluorite	9	corundum
5	apatite	10	diamond

What Did You Learn?

1. Explain how to do a streak test on a rock.
2. Use the Mohs hardness scale to find one mineral that the mineral apatite can scratch and one mineral that it cannot scratch.
3. Name the three big groups of rocks.
4. Which group of rocks contains fossils?
5. Which kind of rocks are formed when igneous rocks or sedimentary rocks are subjected to high temperatures and/or pressure?
6. Which kind of rock forms from the cooling of magma underground before it reaches the surface of the earth?
7. Name three or four sedimentary rocks.
8. Basalt, granite, and obsidian are examples of what kind of rock?
9. Marble is a metamorphic rock that is made from what sedimentary rock under conditions of great heat and pressure?

	Igneous	Sedimentary	Metamorphic
Name Origin	From Latin word *ignis* meaning fire	From the nature of their formation, deposited by wind, water, or glaciers (mostly water)	From the Greek word *meta* meaning change and *morphe* meaning form
Formation	They form as magma cools	Formed as sediment is pressed together often by water	Often formed by being buried, causing temperature and pressure to rise
Varieties / Types	Granite, Basalt, Obsidian, Pumice	Shale, Sandstone, Limestone, Coal	Slate, Gneiss, Quartzite, Marble

WHAT DID YOU LEARN?

1. **Explain how to do a streak test on a rock.** *Rub a rock on a piece of unglazed ceramic tile and observe if it leaves a colored streak. Try to match the color of the streak with what is shown in a reference book or Internet source.*

2. **Use the Mohs hardness scale to find one mineral that the mineral apatite can scratch and one mineral that it cannot scratch.** *Apatite can scratch any mineral with a hardness less than 5 (such as talc, gypsum, calcite, fluorite) and cannot scratch any mineral greater than 5 (such as orthoclase feldspar, quartz, topaz, corundum, diamond).*

3. **Name the three big groups of rocks.** *Sedimentary, igneous, and metamorphic*

4. **Which group of rocks contains fossils?** *Sedimentary*

5. **Which kind of rocks are formed when igneous rocks or sedimentary rocks are subjected to high temperatures and/or pressure?** *Metamorphic*

6. **Which kind of rock forms from the cooling of magma underground before it reaches the surface of the earth?** *Intrusive igneous rocks*

7. **Name three or four sedimentary rocks.** *Sandstone, limestone, shale, conglomerate, and others*

8. **Basalt, granite, and obsidian are examples of what kind of rock?** *Igneous*

9. **Marble is a metamorphic rock that is made from what sedimentary rock under conditions of great heat and pressure?** *Limestone*

INVESTIGATION #14

How Little, Tiny Things Settle Out of Water to Become Rocks

Think about This

Anna's house overlooked the Maple Trail Creek. After several days of hard rains, the usually clear water had become a murky brown and had overflowed the banks. The water kept rising and had gotten close to their house. Anna's dad finally said the floodwaters were starting to recede, but most of the driveway to their house was still covered by sand, brown silt, and twigs. Where do you think these materials came from?

The Investigative Problems

What are the different settling rates (sedimentation rates) of various sizes and types of material?

60

Gather These Things:

✓ Quart jar with lid
✓ Soil
✓ Gravel
✓ Sand
✓ Very small, dry pieces of twigs
✓ Water

Procedure & Observations

1. Put sand, gravel, soil, and a few small dry pieces of twigs in a quart jar until it is about halfway full. Then add water until the jar is about ²/₃ full. You need to leave some room at the top for mixing.

2. Predict which materials you think will settle to the bottom first and which ones will be on top after they have been stirred up.

3. Place the lid on the jar and shake vigorously to mix. Slowly stop shaking the jar, ending with a swirling motion.

4. Place the jar on a flat surface and let it stand until all the materials are settled and the water becomes somewhat clear. It may take several hours for the water to become really clear, but you may begin observing the results after about 15 minutes.

5. Observe and record the order in which the materials settled. Is this what you predicted would occur?

6. Make a drawing of what you observe. Be sure to label the layers that form.

7. Shake the jar again as you did before and place it on a flat surface. Wait 15 minutes.

8. Did you get about the same results that you did the first time?

61

OBJECTIVES

Students should learn that the process of depositing sediment out of water is known as sedimentation. Some particles continue to be carried along while the water is moving but settle out when the water stops moving or slows down.

NOTE

You might want to look for places where water drains across a sandy field or in a shallow stream, so students can observe inside and outside curves in the path of the moving water. Sandy deposits will occur on inside curves, and more erosion will occur along outside curves. Pictures of meandering streams will also show inside and outside curves.

The Science Stuff

You should have noticed that materials were sorted out in order by particle size and composition. The larger, heavier rocks settle out first, followed by sand, and then clay and then finer silt. The very smallest particles may remain suspended for several hours before they finally settle out. Your drawing will probably show a layer of gravel on the bottom of the jar, following by sand, then clay and silt, then humus and other materials that float. The spaces around the rocks will be filled in with sand.

The materials that are deposited by water are known as sediment. The process of depositing sediment out of water is known as sedimentation. As long as the water is moving, some of the particles may be carried along. The finer particles tend to settle out when the water stops or almost stops moving.

When deposits have settled out of water and they become hardened, they are called sedimentary rocks.

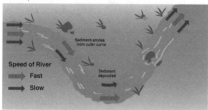

Making Connections

Small streams and rivers carry large amounts of sediment into lakes and oceans. As the water from streams and rivers enters lakes and oceans, the water begins to move much more slowly. The sediment being carried along by the water then starts to settle out. These areas often form deltas. After many years, the delta region can become larger and larger.

Have you ever seen sand bars along one side of a curving stream or river? Sand bars always form along the inside of a curve. That is because the water is moving slowly on the inside of the curve and some of the sand settles out there. The water is moving faster on the outside of the curve where the current keeps carrying sediment along.

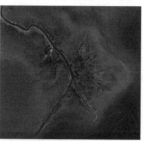

We learned that materials such as stones, sand, clay, and silt can be deposited by water. However, these materials can also be moved and deposited by wind and glaciers. Tons of fine silt and dust can be carried over great distances by winds. Glacier-deposited materials can leave huge rocks and boulders where they were deposited before the glacier began to retreat or melt.

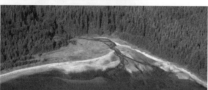

Dig Deeper

- Try to find pictures of sedimentary layers that formed in the area after the Mount St. Helens eruption in 1980. How long did it take for these layers to form? What is the history of these layers?

- Make a model of a river delta. Combine a small amount of sand, gravel, and soil in a jar and shake. Pour the contents of the jar down an incline made from an empty paper towel roll. Cut the roll in half longitudinally and line with aluminum foil. Let the mixture run out onto a flat surface. Describe how the materials are sorted. You may notice that everything doesn't reach the flat surface. Do this a few times and vary the height of the incline. Do some research on river deltas.

- From 1933 to 1942, a number of lakes were created by building dams across small rivers. The purpose of many of these lakes was for flood control, and they were built as part of a federal program known as CCC (Civilian Conservation Corps). See if you can find out more about these programs. What are some of the lakes that were built? (Hint: You may be able to get firsthand information from some of the older people who were alive during this time and remember the CCC programs.)

Construction of the Shasta Dam in California, circa 1942

What Did You Learn?

1. What term refers to materials such as stones, sand, and silt that are deposited by water?
2. What is the process of depositing sediment out of water?
3. What are two other things that can deposit sediments?
4. Is the settling rate faster for larger/heavy materials or for smaller/ lighter materials?
5. Where are sand bars likely to be found — on the inside of a river curve where the water moves slowly or on the outside of a river curve where the water moves faster?
6. The Mississippi River carries tons of sediment into the Gulf of Mexico every year. What is the area called where the river and the gulf meet?
7. Why is most of the sediment deposited in this area rather than farther upstream?
8. What is most likely to leave deposits of large boulders — water, wind, or glaciers?
9. When sediment hardens or consolidates, what kind of rocks are formed?

WHAT DID YOU LEARN?

1. What term refers to materials, such as stones, sand, and silt that are deposited by water?
Sediment

2. What is the process of depositing sediment out of water? *Sedimentation (or deposition)*

3. What are two other things that can deposit sediments? *Wind and glaciers*

4. Is the settling rate faster for large/heavy materials or for small/lighter materials? *Large/heavy*

5. Where are sand bars likely to be found — on the inside of a river curve where the water moves slowly or on the outside of a river curve where the water moves faster? *Inside of a curve*

6. The Mississippi River carries tons of sediment into the Gulf of Mexico every year. What is the area called where the river and the gulf meet? *Delta*

7. Why is most of the sediment deposited in this area rather than farther upstream? *The water is moving very slowly as it enters the Gulf. Moving currents of water are able to carry sediment long distances.*

8. What is most likely to leave deposits of large boulders — water, wind, or glaciers? *Glaciers*

9. When sediment hardens or consolidates, what kind of rocks are formed? *Sedimentary rocks*

How Rocks and Dirt Catch a Ride

Think about This Eli walked outside and called his mom to come quick and look at their back yard. The day before there had been bright red and white blooms on a row of azalea bushes that bordered green grass. Now a thick layer of mud and rocks covered everything. His mom quickly realized what had happened. A night of heavy rain had eroded loose soil that had been piled up by some bulldozers. The land just up the hill from them was being cleared for a new apartment building. "They hadn't counted on the heavy rains causing this much erosion," Mom said, "but the developers have some major cleaning up to do." What do you think erosion had to do with the mess in Eli's back yard?

The Investigative Problems
What are the effects of rain on dirt?

64

Gather These Things:
- ✓ Rectangular pan or cookie sheet
- ✓ Aluminum foil (optional)
- ✓ Paper cup and a toothpick to poke holes in it
- ✓ Water   ✓ Sand or dirt
- ✓ Books   ✓ Paper towels
- ✓ Large pan
- ✓ Small, dry twigs, leaves, pine straw, or mulch

Procedure & Observations

Part A

1. Spread one inch of sand or dirt in the bottom of the pan. (You may want to line the pan with aluminum foil to protect the surface if it has a special coating on it.) Place two books under one end so the pan is slightly tilted. Place the other end of the pan in a plastic container that will catch the overflow of water.

2. Now poke several small holes in the paper cup. Hold the cup over the higher end of the pan and fill it with water. The falling water will act like raindrops hitting the dirt.

3. Observe the effects of the "rain" and record what you see.

4. Refill the cup with water each time it gets empty, and count the number of cups of water it took for all the dirt to be moved to the lower end of the pan. You can move the cup back and forth over the elevated end, but try to keep the distance above the dirt about the same.

5. How many cups of water were needed to move the dirt to the lower end of the pan?

Part B

1. Remove the wet dirt, dry the pans, and repeat the activity. This time add two more books to make the incline sleeper. Repeat the investigation.

2. How many cups of water were needed to move the dirt to the lower end of the pan this time?

3. Did you make any other observations?

4. Remove the wet dirt, dry the pans, and set everything up as in the last activity. This time place small dry twigs, leaves, pine straw, or mulch over the dirt. Repeat the investigation, except allow the same amount of water to fall as in the last activity.

5. What differences did you observe in the amount of erosion this time?

6. Did the twigs, leaves, pine straw, or mulch help to prevent erosion of the soil?

65

OBJECTIVES Students should learn what is meant by erosion and sedimentation and understand some of the ways in which erosion can be prevented.

NOTE If you have time, read a passage from one of Faulkner's books that describes the dry, dusty farmland of the South and discuss what the farmers could have done to take better care of their land. This is a good opportunity to relate this lesson to periods of American history when farmers continued to move farther west to find more fertile soil. Lesson 20 refers to the Dust Bowl of the 1930s. Either lesson can be related to American history.

The Science Stuff

Erosion is a broad term that includes the processes that move soil, sediment, and other materials on the earth from one place to another. Erosion can be caused by water, mudflows, ice, wind, and gravity.

Erosion tends to occur faster on hills than on flat lands. It tends to occur faster on bare ground than on ground that is covered by vegetation. Remember Eli's house? The land up the hill from his house had been cleared of trees and grass to get ready to build a new apartment building. The land was on a hill, and there was nothing to cover the ground, so erosion occurred quickly during a heavy rain.

One of the best ways to stop gully formation and erosion on bare ground is to plant trees and grass in the ground. Serious erosion can occur in a short time on bare, hilly ground.

When rain falls, it can either be absorbed by the land or it will become runoff. As the water travels over the land as runoff, it moves particles from the land and causes erosion. Runoff water, with its eroded particles, goes into drainage ditches, streams, and rivers. Eventually it goes into large rivers that empty into oceans. Large amounts of the eroded particles get deposited in the delta areas (where a river empties into an ocean) as the river water slows its movement.

Topsoil is a valuable resource in any location where it is found, but it can easily be lost by unwise uses of the land. It has been estimated to take hundreds of years for the combined action of plant growth, bacterial decay, and erosion action to produce a few centimeters (about an inch) of good topsoil.

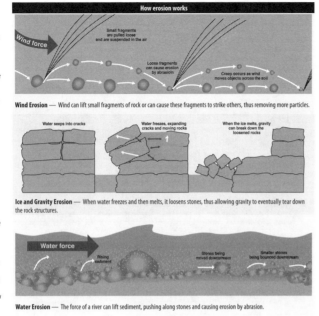

How erosion works

Wind Erosion — Wind can lift small fragments of rock or can cause these fragments to strike others, thus removing more particles.

Ice and Gravity Erosion — When water freezes and then melts, it loosens stones, thus allowing gravity to eventually tear down the rock structures.

Water Erosion — The force of a river can lift sediment, pushing along stones and causing erosion by abrasion.

66

Making Connections

William Faulkner is a famous author. Some of his stories were about poor hill farmers who couldn't produce enough crops to live on. They didn't understand that after years of plowing their fields on hilly slopes, erosion had carried away much of the topsoil and carved huge gullies on their land. Although his purpose in writing wasn't to teach his readers about conservation of topsoil, his stories certainly describe what happens when the land is no longer able to produce good crops.

One of the techniques recommended by soil conservation groups is a practice known as contour plowing. Fields are plowed along the curve of a hill instead of up and down the hill. Contour plowing helps slow the runoff from a rain.

As you learned in the previous lesson, rivers that flow across a flood plain almost always begin to meander or form curves. The water that goes around the outside of the curve is moving faster, so there is erosion of the riverbank. The slower-moving water on the inside of the curve allows sand to settle out and be deposited. Eventually the curves will touch and the water will make a new pathway, leaving an oxbow body of water.

The Mississippi River with oxbow lakes.

Formation of Oxbow Lakes

A Narrow neck of the meander is gradually being eroded.

B Deposition takes place, sealing off the old meander.

C Water now takes the quickest route and the meander neck has been cut through completely.

D Oxbow lake left behind when meanders are completely cut-off.

Dig Deeper

- Gullies tend to form in many areas where the topsoil has been eroded. Find some of the ways in which gully formation can be slowed or stopped.

- Find a map of the drainage system for where you live. See if you can trace the path of rainwater that runs off your street or road until it reaches the ocean.

What Did You Learn?

1. What is erosion?

2. What things can cause erosion?

3. How are erosion and sedimentation different?

4. Under what conditions will erosion occur at a faster rate?

5. How can erosion be slowed or stopped?

6. Why is it important to take care of topsoil in an area?

7. Use diagrams to show how a meandering river can change course after several years.

8. What is an oxbow lake?

67

WHAT DID YOU LEARN?

1. **What is erosion?** *Erosion is a broad term that includes the processes that move soil, sediment, and other materials on the earth from one place to another.*

2. **What things can cause erosion?** *Erosion can be caused by water, mudflows, ice, wind, and gravity.*

3. **How are erosion and sedimentation different?** *Erosion is a process that moves soil and other materials away and sedimentation is a process that leaves deposits of sediment behind.*

4. **Under what conditions will erosion occur at a faster rate?** *On hills and on bare ground*

5. **How can erosion be slowed or stopped?** *By covering ground with vegetation, especially on hilly areas*

6. **Why is it important to take care of topsoil in an area?** *It may take hundreds of years to produce a good, thick layer of topsoil, and it can be lost in a few years by erosion.*

7. **Use diagrams to show how a meandering river can change course after several years.** (See text page 67)

8. **What is an oxbow lake?** *When the curves of a meandering river meet and produce a new pathway for the river, the old river curve will remain as an oxbow lake.*

Physical and Chemical Weathering

Think about This
Kody's mom squealed the brakes as she suddenly stopped the car. Kody looked up to see a landslide right in front of them that was blocking the road. Mom quickly made a call to the highway department and reported what had happened. "It looks like we're going to have to turn around and take the long way home again," she said.

This was the second time in a month a landslide had blocked the road. "I wonder how long it would take for the whole mountain to fall down," Kody wondered.

Do you think mountains eventually wear down and get smaller?

Procedure & Observations

Part A. Physical Weathering
1. Fill an aluminum soft drink can as full of cold water as you can. Place it in a container with a flat surface and put it in the freezer. Leave it for several hours until the water is frozen solid.
2. Describe what happened to the aluminum can.
3. Try to give an explanation for what happened before you read "The Science Stuff."
4. Compare your explanation with the explanation in the book.

The Investigative Problems
How do rocks get worn down?

68

Gather These Things:
✓ Empty aluminum soft drink can
✓ Disposable, plastic quart-size jar with screw-on lid
✓ Water   ✓ Small margarine tubs
✓ Salt   ✓ Vinegar
✓ Handful of rocks from outside
✓ Several pieces of hard candy (rounded or cylindrical, at least 1 cm thick with no outer coating)
✓ Limestone rocks (garden or landscaping store)
✓ Steel wool pad (remove any soap)

5. Combine a handful of rocks in a disposable plastic jar with some hard candy. Screw the lid on the jar and shake the jar so that the rocks and candy alternately hit the lid and the bottom of the jar. Do this several times. Stop shaking when you see several broken pieces of candy or rocks. Examine the contents.
6. Look carefully at the broken fragments. Describe how they look. Draw a few of the fragments.
7. Compare the broken fragments with the weathered rocks. Which fragments had the most sharp, angular shapes?
8. Were there more fragments of candy or more fragments of rocks?
9. Put one or two drops of water on some of the candy fragments that have clumped together and let dry. This represents rock pieces that get cemented together.
10. Would cementing rock pieces together be an example of weathering or would it be the opposite of weathering?

Part B. Chemical Weathering
1. Take some plain steel wool (which is made of iron) and dip it in a solution of salt water. Place the pad in an empty margarine tub or other container. Set it aside but continue to dip it in the salt water from time to time. After a few days, examine the steel wool again.
2. What color do you observe that was not present when you began this investigation?
3. Is the new substance hard, or is it crumbly?

Optional
1. You may want to repeat the investigation in which you put limestone rocks in a container with vinegar. This represents another example of chemical weathering by exposure to an acid.
2. Describe what happens.
3. Describe the limestone rocks after you have removed them from the vinegar and washed them in water.

69

OBJECTIVES Students should learn the differences in physical and chemical weathering processes.

NOTE Remind students that erosion can occur very rapidly or very slowly.

Help them think of examples of both.

The Science Stuff

Weathering processes are constantly breaking down rocks, as well as houses, cars, and anything else exposed to weathering agents. Physical weathering causes rocks to break down into small pieces or to become more rounded.

Shaking the jar of rocks and hard candy was like physical weathering. You should have observed that the hardest rocks were the least likely to break when shaken together. Pieces of the candy and rocks that do break will have sharp, jagged or angular shapes. Although the rocks and candy were shaken and broken, the chemical properties and chemical make-up of the rocks and the candy did not change. The picture on page 68 shows freshly broken jagged rocks. These rocks will become more rounded after several years of weathering.

In nature, continual physical weathering by water, wind, ice, or grinding against other rocks causes sharp edges to become more rounded. Whenever you find very smooth, rounded rocks, you know they have undergone more weathering than rocks with sharp edges.

Changing temperatures can also cause physical weathering. Most objects expand when they are heated and contract when they are cooled. The freezing of water is an important exception to this principle, because water expands when it freezes. When water gets into cracks in rocks and freezes, the rocks often break. You can get an idea of how powerful frozen water can be as you examine your frozen soft drink can.

Chemical weathering is another way in which rocks are broken down. Chemical weathering involves chemical changes that produce new substances. You may remember putting limestone in vinegar (an acid). Bubbles being released from the limestone indicated that new substances were forming. Chemical weathering in nature doesn't usually involve vinegar, but it does often involve another acid known as carbonic acid. This acid forms when carbon dioxide from the air reacts with water vapor. If sulfur or nitrogen oxides (both are pollutants) are released into the air, they may also react with water to form much stronger acids. Any kind of acid will cause chemical weathering of limestone rocks.

Iron is found in many rocks, but it usually exists as an iron oxide. The reddish color of many rocks is an iron oxide. Although iron metal is very hard, iron oxide is not. When iron oxide forms in rocks, it tends to crumble and may cause nearby rocks to fall as well.

Physical and chemical weathering processes may cause very slow changes in rocks and minerals. They may also cause sudden dramatic changes, such as the rockslide mentioned in this lesson. In a landslide, a large amount of rocks, soil, and plants breaks loose and falls down the mountain, and often blocks the highway. Some rocks may seem like they would last forever, but sooner or later, they tend to give way to physical and chemical weathering.

The steel of the Golden Gate Bridge in San Francisco, California, must be continually touched up with acrylic paint to protect the metal from corrosion.

70

Making Connections

Highways also undergo physical and chemical weathering, causing potholes to form in the road. In places where temperatures get below freezing during winter months, water can get into potholes and freeze. Remember what happened to the aluminum can as water froze inside it and you can get a good idea of what happens to potholes in highways that contain frozen water.

Iron supports in bridges have to be treated with special coatings and paints to prevent rust from forming and weakening the bridges. Iron metal is a strong substance that can support heavy weight, but iron oxide is weak and crumbly.

Dig Deeper

- Pour Plaster of Paris in a margarine tub until about half full. Place some lima bean seeds in while the plaster is still in a liquid state. Let harden. Cover with a moist paper towel. Leave it in a warm place for a few days. Remoisten the paper towel every few days until the seeds sprout. Describe what happens to the hardened plaster. Do some research about how plants play a role in weathering rocks.

- You can buy a rock tumbler that constantly rolls rocks around inside of them. After leaving rocks in the tumbler for several days, you may have some very attractive smooth stones. Even jagged pieces of rocks will eventually become rounded.

What Did You Learn?

1. Suppose a rock breaks into several small pieces after a big rock falls on it. Is this an example of physical weathering or chemical weathering? Explain your answer.

2. What kind of shape would you expect newly broken rocks to have — sharp and jagged or smooth and rounded?

3. Give some examples of how rocks can undergo physical weathering.

4. Give some examples of how rocks can undergo chemical weathering.

5. Why do rocks sometimes break during freezing weather when water gets into cracks in the rocks?

6. What are some of the differences between iron and iron oxide?

Pause and Think

The ultimate example of sedimentation and erosion is the Grand Canyon. Read the narrative story at www.investigatethepossibilities.org, for a creationist view of the Canyon.

71

WHAT DID YOU LEARN?

1. Suppose a rock breaks into several small pieces after a big rock falls on it. Is this an example of physical weathering or chemical weathering? Explain your answer. *This is physical weathering, because the pieces of rock all have many of the same properties they had before they broke.*

2. What kind of shape would you expect newly broken rocks to have — sharp and jagged or smooth and rounded? *Sharp and jagged*

3. Give some examples of how rocks can undergo physical weathering. *Exposure to moving water, wind, ice; grinding against other rocks; changing temperatures cause rocks to expand when they are heated and contract when they are cooled; freezing of water in cracks in rocks; growth of plants; landslides.*

4. Give some examples of how rocks can undergo chemical weathering. *Rusting of rocks that contain iron; reaction of certain rocks with acids; other chemical reactions of chemicals in rocks.*

5. Why do rocks sometimes break during freezing weather when water gets into cracks in the rocks? *Water expands when it freezes and exerts pressure on the surrounding rocks.*

6. What are some of the differences between iron and iron oxide? *Iron is a strong, dark-colored, shiny metal. Iron oxide is a reddish-colored, dull material that crumbles easily.*

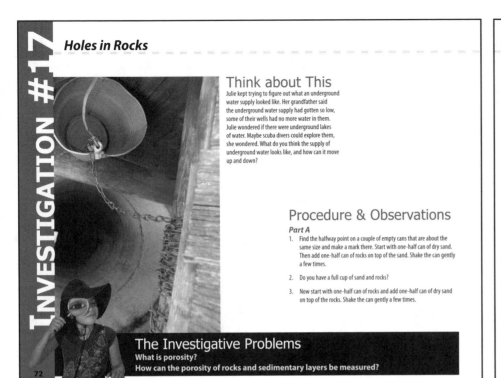

INVESTIGATION #17

Holes in Rocks

Think about This

Julie kept trying to figure out what an underground water supply looked like. Her grandfather said the underground water supply had gotten so low, some of their wells had no more water in them. Julie wondered if there were underground lakes of water. Maybe scuba divers could explore them, she wondered. What do you think the supply of underground water looks like, and how can it move up and down?

Procedure & Observations

Part A

1. Find the halfway point on a couple of empty cans that are about the same size and make a mark there. Start with one-half can of dry sand. Then add one-half can of rocks on top of the sand. Shake the can gently a few times.

2. Do you have a full cup of sand and rocks?

3. Now start with one-half can of rocks and add one-half can of dry sand on top of the rocks. Shake the can gently a few times.

The Investigative Problems

What is porosity?
How can the porosity of rocks and sedimentary layers be measured?

72

Gather These Things:

✓ Six clean empty tin cans, same size, with lids removed
✓ Container about three or four times larger than the paper cup
✓ Dry sand   ✓ Wire strainer
✓ Small to medium size rocks
✓ Metric liquid measuring cup
✓ Small paper cup
✓ Sturdy toothpick

4. Do you have a full cup of sand and rocks?

5. Try to think of a reason for the difference.

6. Pour the rocks and sand through a wire strainer to separate them to continue using them.

7. Fill a can with water and pour the water into a metric measuring cup.

8. Record the volume of the water in milliliters that an empty can will hold.

Part B

1. Fill one can to the top with rocks. Fill another can to the top with water. Carefully pour as much of the water as possible over the rocks without letting the water spill over. Measure the water that is left in the can. Subtract from the volume of the full can to find how much water went into the spaces around the rocks.

2. Record the amount of water that went into the spaces around the rocks in milliliters.

3. Fill a can to the top with dry sand. Slowly add a can full of water to the sand without letting the water spill over. Measure the water that is left in the can. Subtract from the volume of the full can to find how much water went into the spaces around the particles of sand.

4. Record the amount of water that went into the spaces around the sand in milliliters.

5. Compare the amount of water that went into the spaces around the rocks with the amount that filled in around the sand.

6. Which had the greater porosity — a can of rocks or a can of sand?

7. Punch holes into a paper cup with a toothpick. Place the paper cup in a bowl that is three or four times larger than the paper cup. Place dry sand in the container around the paper cup. Slowly add one-half can of water to the sand. After a few minutes, observe what happens inside the paper cup.

8. Record your observations.

73

OBJECTIVES

Students should learn that rocks and soils have spaces in them that can be filled with water, oil, or smaller particles. Porous soil allows water to be absorbed and reach the roots of plants. Without underground porous rocks that allow water to move through the spaces, there would be no water table or source of clean water.

NOTE

If you would like for your students to work on some measuring skills, add the following investigations.

Fill an empty 16-ounce plastic soda bottle completely full of water. Screw the top on tightly. Measure the distance around the middle of the bottle to the nearest millimeter.

1. What is the distance around the middle of the bottle to the nearest millimeter?

Now freeze the bottle. Remove from the freezer and loosen the cap. Measure the distance around the middle of the bottle again to the nearest millimeter. The bottle may break, but it will most likely just expand around the middle.

2. What is the distance around the middle of the bottle after it was frozen to the nearest millimeter?

3. Try to give an explanation for why the distance around the bottle increased after the water was frozen.

If you have a sensitive set of scales, place some limestone rocks in a container of vinegar, weigh the container, rocks and vinegar as soon as you can.

1. Weigh the container, limestone rocks, and vinegar as soon as you can. Then weigh them again after a few days.

The Science Stuff

There are spaces around rocks, sand, and other particles. These spaces can be filled with anything that is small enough to fit in them. Water can go into the spaces around rocks, sand, and other earth materials.

The spaces around rocks, sand, or other earth materials are called pores. The amount of spaces compared to the total volume of the material is its porosity. Notice that a container of rocks has more porosity than a container of sand.

An aquifer is a rock formation with enough porosity to hold water and allow water to be moved in and out of the rock. The paper cup you punched holes in illustrates how water goes into a well from an aquifer. When the pores are full of water, some of the water will move into the cup.

The water table is the underground area that is saturated with water. Next to a river or lake, the top of the water table is the same as the surface of the river or lake.

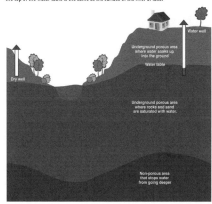

Making Connections

When geologists are looking for underground oil deposits, they are actually looking for layers of rock that have a high porosity. Some deep rock layers contain oil or gas or both. These deposits are found in the spaces in the rock layers just as water occupies spaces in water aquifers. There is usually some water in oil and gas deposits as well. Sometimes oil and gas deposits are thousands of meters (over a mile) deep.

The amount of pressure associated with oil deposits may be an important clue about the age of the deposits. The following personal account was provided by Jill Whitlock (formerly with CSI). "When I was working as an exploration geologist, a friend of mine was sitting on a well in Louisiana that was drilling to 12,000 feet. About 10,000 feet they hit an over-pressured zone and blew 10,000 feet of drill pipe out of the hole. That much pressure still in the reservoir rock indicates that it is very young. The pressure bleeds off over a few thousand years and will not be maintained for long ages. Therefore, over-pressured zones are a very good indicator that the earth and its rocks are young."

A group of mission partners from the United States are helping people in some of the West African coastal countries to drill for clean water. The coastal land is just above sea level and is flat. People often drink from streams and rivers. Most of their local wells are shallow, because they don't have to dig very deep to reach the water table. However, both the shallow wells and the streams of water often test positive for pollutants. There is much cleaner water deep underground, but in order to reach this source of water, special drilling equipment must be used. Pumps must then be used to get the water to the surface. Clean water is taken for granted in the United States, but in some of the poor countries of the world, it is a luxury.

Dig Deeper

- Do some research and see if you can find more information about how oil and gas deposits were made in the first place. Remember, there will be completely different explanations from scientists who believe in a worldwide flood and in scientists who do not. Try to find explanations from both groups.

- Do some research and see if you can find more information about how oil and gas deposits are located and then brought up to the surface.

- Even today in some places, farmers dig their own wells, but 50 or 60 years ago in the United States it was common for farmers to dig their own water wells by removing one bucketful of soil at a time. The process involved hard work and was dangerous. Try to find an older person who remembers how wells were dug and see if you can get an interview. You might be able to find something on the Internet as well. Write a report on what you learn.

A typical pump jack oil well

What Did You Learn?

1. When farmers dig wells in order to get water to drink, how deep do the wells have to be in order to reach water?

2. What might cause a water well to go dry?

3. The spaces between rocks, sand, or other earth materials are called what?

4. The amount of space between rocks, sand, or other earth materials compared to the total volume of the material is known as what?

5. What is a water aquifer?

6. Are oil and gas deposits found in porous or nonporous layers?

7. Are most oil and gas deposits found near the surface of the ground or deep below the ground?

8. When water has filled all the pores in a porous underground area, the area is said to be _____ with water.

9. Why is there a great need for people all over the world to find drinking water from clean wells rather than getting their drinking water from streams and shallow polluted wells?

74

75

WHAT DID YOU LEARN?

1. When farmers dig wells in order to get water to drink, how deep do the wells have to be in order to reach water? *Wells must reach the water table, the place where water has saturated a layer of rocks, sand, and other soil material.*

2. What might cause a water well to go dry? *The water table might become lower due to insufficient rain.*

3. The spaces between rocks, sand, or other earth materials are called what? *Pores*

4. The amount of spaces between rocks, sand, or other earth materials compared to the total volume of the material is known as what? *Porosity*

5. What is a water aquifer? *A rock formation with enough porosity to hold water and allow water to be moved in and out of the rock.*

6. Are oil and gas deposits found in porous or nonporous layers? *Porous layers*

7. Are most oil and gas deposits found near the surface of the ground or deep below the ground? *Deep below the ground*

8. When water has filled all the pores in a porous underground area, the area is said to be ____ with water. *Saturated*

9. Why is there a great need for people all over the world to find drinking water from clean wells rather than getting their drinking water from streams and shallow, polluted wells? *In some places around the world, streams and shallow wells are dangerously polluted with disease-causing organisms, but are the only available source of drinking water. Water from clean wells would greatly improve the health of people in these areas*

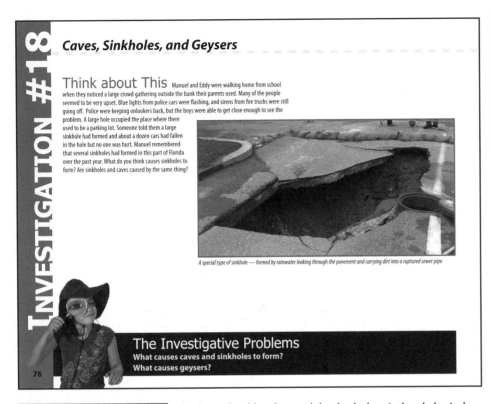

INVESTIGATION #18

Caves, Sinkholes, and Geysers

Think about This

Manuel and Eddy were walking home from school when they noticed a large crowd gathering outside the bank their parents used. Many of the people seemed to be very upset. Blue lights from police cars were flashing, and sirens from fire trucks were still going off. Police were keeping onlookers back, but the boys were able to get close enough to see the problem. A large hole occupied the place where there used to be a parking lot. Someone told them a large sinkhole had formed and about a dozen cars had fallen in the hole but no one was hurt. Manuel remembered that several sinkholes had formed in this part of Florida over the past year. What do you think causes sinkholes to form? Are sinkholes and caves caused by the same thing?

A special type of sinkhole — formed by rainwater leaking through the pavement and carrying dirt into a ruptured sewer pipe

The Investigative Problems

What causes caves and sinkholes to form?
What causes geysers?

76

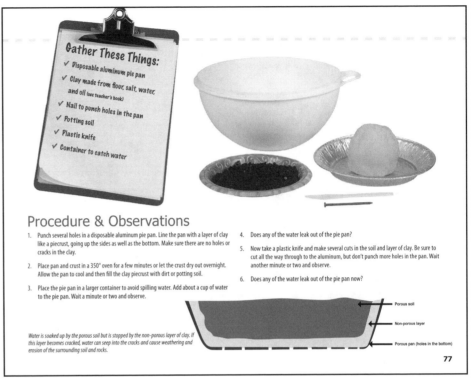

Gather These Things:

✓ Disposable aluminum pie pan
✓ Clay made from flour, salt, water, and oil (see teacher's book)
✓ Nail to punch holes in the pan
✓ Potting soil
✓ Plastic knife
✓ Container to catch water

Procedure & Observations

1. Punch several holes in a disposable aluminum pie pan. Line the pan with a layer of clay like a piecrust, going up the sides as well as the bottom. Make sure there are no holes or cracks in the clay.

2. Place pan and crust in a 350° oven for a few minutes or let the crust dry out overnight. Allow the pan to cool and then fill the clay piecrust with dirt or potting soil.

3. Place the pie pan in a larger container to avoid spilling water. Add about a cup of water to the pie pan. Wait a minute or two and observe.

4. Does any of the water leak out of the pie pan?

5. Now take a plastic knife and make several cuts in the soil and layer of clay. Be sure to cut all the way through to the aluminum, but don't punch more holes in the pan. Wait another minute or two and observe.

6. Does any of the water leak out of the pie pan now?

Water is soaked up by the porous soil but is stopped by the non-porous layer of clay. If this layer becomes cracked, water can seep into the cracks and cause weathering and erosion of the surrounding soil and rocks.

— Porous soil
— Non-porous layer
— Porous pan (holes in the bottom)

77

OBJECTIVES

Students should understand that both chemical and physical weathering, as well as erosion, can occur underground.

Underground weathering and erosion may result in caves and sinkholes. Water often moves from place to place through cracks in rocks.

NOTE

Start with a few cuts in the clay. Add more cuts until the water begins to leak out. These cuts represent cracks in rocks. The recipe for clay in lesson 8 can be used here, but you may need to bake the pan a few minutes to dry out the clay before making cracks.

The Science Stuff
We have investigated ways in which rocks are weathered and eroded. Rock particles can be transported from one place to another. It's easy to see the effects of weathering and erosion on the earth's surface. However, that's not the only place where weathering and erosion can occur. Weathering can also occur underground.

Underground water can cause erosion just as it does on the earth's surface. For example, in the Kentucky area, there are hundreds of miles of interconnected caves. Many of these caves were formed from weathering and erosion of limestone rock (calcium carbonate) as water moved long distances through cracks in the limestone. Both physical and chemical weathering and erosion cause these caves to form.

If you have ever been in a cave, you probably remember that it was wet. There may have been a stream or lake in the cave, and there was probably water dripping from the ceiling. If the rock formations around the cave are made of limestone, calcium carbonate is present. As the water drops from the ceiling, small amounts of calcium carbonate are dissolved and deposited. These deposits can form beautiful structures known as stalactites that hang from the ceiling and stalagmites that rise from the ground.

Since these structures form drop by drop, it is often assumed that large caves took millions of years to form. However, the basement below the Lincoln Memorial, and some other places, contain both stalactites and stalagmites caused by drop-by-drop deposits that came from weathering and erosion of the stone structures above. Many geologists now think large cave formations could have formed in hundreds or thousands of years.

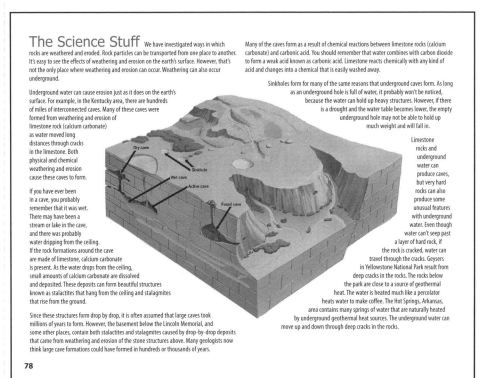

Many of the caves form as a result of chemical reactions between limestone rocks (calcium carbonate) and carbonic acid. You should remember that water combines with carbon dioxide to form a weak acid known as carbonic acid. Limestone reacts chemically with any kind of acid and changes into a chemical that is easily washed away.

Sinkholes form for many of the same reasons that underground caves form. As long as an underground hole is full of water, it probably won't be noticed, because the water can hold up heavy structures. However, if there is a drought and the water table becomes lower, the empty underground hole may not be able to hold up much weight and will fall in.

Limestone rocks and underground water can produce caves, but very hard rocks can also produce some unusual features with underground water. Even though water can't seep past a layer of hard rock, if the rock is cracked, water can travel through the cracks. Geysers in Yellowstone National Park result from deep cracks in the rocks. The rocks below the park are close to a source of geothermal heat. The water is heated much like a percolator heats water to make coffee. The Hot Springs, Arkansas, area contains many springs of water that are naturally heated by underground geothermal heat sources. The underground water can move up and down through deep cracks in the rocks.

Making Connections
There are many geysers in Yellowstone National Park. One of the most famous geysers is known as Old Faithful, because it spews out hot water at very predictable times. The water seems to move through deep cracks on the ground, and is often compared to how a coffee percolator works.

There is also a great deal of underground water from a variety of sources. When a volcano erupts, steam is almost always present. Geysers spew out steam and water as they erupt. Deep sources of water can sometimes be reached by drilling beyond the ordinary water table.

Dig Deeper
- Use a map to show the locations of natural caves in the United States. What kinds of rocks are found in these places?
- Compare the different kinds of caves in the United States.

What Did You Learn?
1. When there is a rain, part of the water becomes runoff. What happens to the rest of the water?
2. What determines how deep water soaks into the ground?
3. Give an example of how underground rocks can be weathered physically.
4. Give an example of how underground rocks can be weathered chemically.
5. Briefly explain how stalactites and stalagmites form in caves.
6. Is the water that erupts from a geyser in Yellowstone National Park hot or cold?
7. Where does the water come from that is spewed out of a geyser?
8. When is a sinkhole most likely to fall in — when there has been plenty of rain or during a drought?

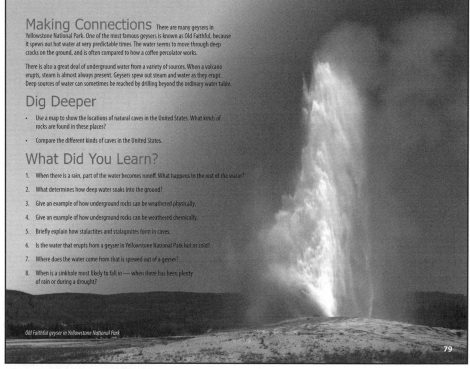
Old Faithful geyser in Yellowstone National Park

WHAT DID YOU LEARN?

1. When there is rain, part of the water becomes runoff. What happens to the rest of the water? *It is soaked up into the ground.*

2. What determines how deep water soaks into the ground? *The porosity of the soil. If there is a layer of soil, such as clay, that is not porous, then water cannot get past that layer (unless there are cracks in the clay layer). As long as the soil is porous (or cracked), it can seep farther down.*

3. Give an example of how underground rocks can be weathered physically. *Even though water can't seep past a layer of hard rock, if the rock is cracked, water can travel through the cracks and cause a wearing down of the rocks.*

4. Give an example of how underground rocks can be weathered chemically. *When water combines with carbon dioxide, a weak acid known as carbonic acid forms. Carbonic acid reacts chemically with limestone rocks, causing the limestone to come apart.*

5. Briefly explain how stalactites and stalagmites form in caves. *As the water drops from the ceiling, small amounts of calcium carbonate are deposited. These deposits form slowly from the ceiling as stalactites or rise from the floor as stalagmites.*

6. Is the water that erupts from a geyser in Yellowstone National Park hot or cold? *Hot*

7. Where does the water come from that is spewed out of a geyser? *It moves up through cracks in the rocks from deep below the surface near a source of magma.*

8. When is a sinkhole most likely to fall in — when there has been plenty of rain or during a drought? *During a drought*

Glaciers

Think about This

Glaciers often leave piles of rocks and soil behind after they melt or "retreat." Sometimes the rocks left by a glacier are very large and heavy. Do you think the glaciers pushed the rocks and soil like a giant bulldozer? Do you think the rocks and soil are carried in a different way?

The Investigative Problems
How do glaciers transport rocks and other materials?

80

Gather These Things:
- ✓ Small plastic container
- ✓ Medium plastic container
- ✓ Large flat container
- ✓ Water
- ✓ Sand and dirt
- ✓ Small rocks
- ✓ Paper towels

Procedure & Observations

Part A

1. Pour about three centimeters (just over an inch) of water into a small plastic container. Place the container in a freezer until the water is frozen.

2. Run warm water over the bottom of the container to loosen the block of ice. Sprinkle a little sand or dirt over the ice. Place some small rocks in another plastic container that is a little larger than the block of ice. Place the block of ice over the rocks, and place a few more rocks on top of the ice.

3. Wait about three minutes and then return the ice and rocks to the freezer. Wait at least 30 minutes before examining the rocks and ice.

4. Predict what you think will happen to the rocks.

5. Remove the ice and rocks from the freezer and place them on a paper towel. Make careful observations of what you see.

6. Did the rocks underneath the ice stick to the ice?

7. Did the rocks seem to be attached to the ice?

8. What did you observe about the rocks on top of the ice?

9. Make a diagram showing how the rocks were attached to the ice.

10. Was this what you predicted would happen?

Part B

1. Now put the rocks and ice in one end of a large flat container. Tilt the container a few degrees and let the ice melt.

2. Describe what you observe after the ice has melted.

3. Make a hypothesis about what the rocks would do to the land if this were a large, heavy glacier that was moving.

81

OBJECTIVES

Students should learn that glaciers can scour and erode the surface of the ground in major ways. They do this mainly by processes known as plucking and abrasion.

NOTE

Most students have misconceptions about glaciers. They tend to think that glaciers operate like a bulldozer and push soil in front of them, but that is not what usually happens. You might want to begin the lesson by having students tell what they think happens when glaciers move and then have them tell you again after they complete the lesson.

The Science Stuff

As glaciers move, they erode, transport, and deposit rocks and other materials. Glaciers change the land in two major ways: by plucking and by abrasion.

Glacial movement is normally very slow, but sudden slippages can result in rapid surges in movement. This is similar to how rocks moving along a fault can lock up for a while and then suddenly break free.

In real glaciers there may be large amounts of dirt, sand, rocks, and plant material attached to the ice, especially on the bottom. Over the years, a huge amount of rock material may be transported from one place to another.

From time to time, glaciers will melt a little bit. Geologists believe that glaciers actually float on melted water as they move to lower elevations. As temperatures get colder, the rocks will freeze or refreeze to the glacier. You should have observed from your investigation that the rocks were firmly attached to the ice after a period of refreezing.

Melted glacier water seeps down into cracks in the rocks below the glacier. As the water refreezes, it expands the cracks, eventually breaking off pieces of rocks. Some of these broken pieces of rock may freeze to the bottom of the glacier. A slow-moving glacier may transport large boulders or broken chunks of rock for long distances. The process by which a glacier picks up rocks and transports them is known as plucking. Plucking is the main way in which glaciers move rocks and soil rather than pushing them like a bulldozer.

Glaciers are also a source of erosion known as abrasion. The jagged rocks that are attached underneath the glacier grind and scrape the land much like sandpaper rubs a piece of wood. By the time the glacier stops moving, the plucked rocks are also somewhat worn down and smoothed off.

When a glacier begins to melt, the rocks and other materials that were frozen to it are left behind as glacial deposition. Till deposits are one kind of glacial deposit. They are unsorted and unstratified deposits containing both large and small boulders. A terminal moraine forms at the end of a glacier when sediment is deposited as the ice melts. These deposits sometimes form a natural dam and create a lake. In your investigation, the rocks that were left after the ice melted represent glacial deposits.

There are two types of glaciers — valley glaciers and continental glaciers. Valley glaciers form in and move down valleys to seek sea level. Continental glaciers are in the form of huge sheets of ice that move over plains and low, hilly land.

Types of Glaciers

Valley Glacier

Continental Glacier

Dig Deeper

- Valley glaciers have produced some of the most scenic places on earth. Find pictures of U-shaped valleys, horns, hanging valleys, truncated spurs, cirques, moraines, and other features that were produced by valley glaciers. Give the location of each picture you find.

- A great deal of the earth's water is in the form of glaciers. Potentially, during an ice age, more of the ocean water becomes frozen as glaciers, and the sea level can get lower. On the other hand, it the earth gets warmer, more glaciers melt and the sea level can get higher. There have been a number of interesting articles written about the possibility of glaciers melting (or not melting). Find an article from the recommended Internet sources and write a report on this.

Making Connections

During the Ice Age, just over 4,000 years ago, continental ice sheets moved over thousands of square miles in North America and other parts of the world. In America, evidence of glacier movement is widespread in the Great Lakes areas (shown here) and in the Dakotas.

What Did You Learn?

1. How do glaciers "pluck" rocks as they move?

2. Are glaciers able to transport rocks and other materials from place to place?

3. Are glacial deposits primarily pushed ahead of moving glaciers or moved by the process of plucking?

4. Are till deposits and moraines laid down by glaciers, wind, or water?

5. Briefly describe the two kinds of glaciers.

6. Explain how glaciers cause erosion (or abrasion).

7. Why do rocks attached to glaciers start out jagged and end up being worn down and rounded?

WHAT DID YOU LEARN?

1. **How do glaciers "pluck" rocks as they move?** *Melted glacier water seeps down into cracks in the rocks below the glacier. As the water refreezes, it expands the cracks, eventually breaking off pieces of rocks. Some of these broken pieces of rock may freeze to the bottom of the glacier.*

2. **Are glaciers able to transport rocks and other materials from place to place?** *Yes*

3. **Are glacial deposits primarily pushed ahead of moving glaciers or moved by the process of plucking?** *Moved by the process of plucking*

4. **Are till deposits and moraines laid down by glaciers, wind, or water?** *By glaciers*

5. **Briefly describe the two kinds of glaciers.** *Valley glaciers form in and move down valleys to seek sea level. Continental glaciers are in the form of huge sheets of ice that move over plains and low, hilly land.*

6. **Explain how glaciers cause erosion (or abrasion).** *The jagged rocks that are attached underneath the glacier grind and scrape the land much like sandpaper rubs a piece of wood.*

7. **Why do rocks attached to glaciers start out jagged and end up being worn down and rounded?** *The newly broken rocks that are plucked up by a moving glacier tend to be jagged, but as they grind and rub against the land, they get worn down.*

Toiling in the Soil

Think about This

Gerri was visiting her grandparents, as she often did. There were an abundance of trees around the house that had been many shades of gold, red, and brown a few weeks earlier. Now most of the leaves had fallen off. Her grandfather told her to look at a clump near the top of one of the trees, but she didn't see anything unusual about it. "That's a squirrel nest," he pointed out. "I've seen a squirrel going in and out of it all morning." Gerri had played in these woods since she was a little girl. She had seen plenty of squirrels but had never noticed their nests before. Now that the leaves were off the trees, they were able to see what looked like four different nests from where she was standing. Have you ever gone to the same place for a long time and then one day noticed something you had never seen before?

The Investigative Problems

What's in the soil we walk on every day?

84

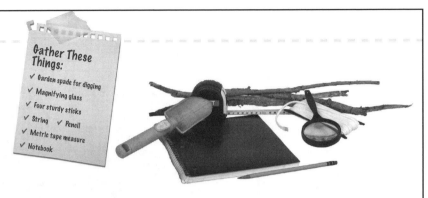

Gather These Things:

✓ Garden spade for digging
✓ Magnifying glass
✓ Four sturdy sticks
✓ String   ✓ Pencil
✓ Metric tape measure
✓ Notebook

Procedure & Observations

Part A

1. Your teacher will show you a plot of ground that contains some grass or other vegetation for you to investigate. (Be sure you have teacher approval before you start digging in an area.)

2. Mark off an area that measures about 30 centimeters (about a foot) by 30 centimeters. (You can mark off a larger area if your teacher agrees.) Outline your area by pushing four sticks into the ground at the corners and connecting them with a piece of string.

3. Carefully observe everything you can see in the area. Notice the kinds of plants there are and any insects or other living things that are there. Observe the rocks, soil, leaves, and twigs on the ground, and anything else you see.

4. Describe or draw what you see.

5. Use the garden spade to remove the grass, plants, rocks, and other objects on the top of the ground. Then start at one corner and begin digging into the soil. Systematically dig a few centimeters (about an inch) into the ground over your area.

6. Make another list of things you find as you dig in the soil. Look for worms, insects, roots, rocks, and other things.

7. Describe or draw what you see as you dig into the soil.

8. Dig up a spade full of soil and hold it in your hands. Observe how it feels, smells, and looks. Notice if it is moist or dry.

9. Try to make the soil into a ball. See if it sticks together or remains loose. Does it feel like sand, like dust, or like something else?

10. Describe the soil you dug up.

Part B

1. Now go back and look at the things you found with a magnifying glass. Add to your notes anything you didn't already notice without using the magnifying glass.

2. What other information can you add to your notes?

3. See if you can find the names of any of the things in your notes.

4. When you are finished, try to put the soil and plants back like you found them.

85

OBJECTIVES

Soil conservation is a very important idea. Soil is one of a nation's major resources and needs to be protected and conserved.

NOTE

There are many ways in which this lesson can be connected to history and literature. You may be able to find a good documentary or some firsthand accounts about the Dust Bowl of the 1930s. Try to help students understand what a valuable resource topsoil is and how it can be protected and preserved.

The Science Stuff

The top few centimeters (an inch or so) of the earth are one of the world's most valuable resources. An abundance of good soil is essential for crops of food that feed the world.

Soil forms slowly over a period of time as big rocks get broken into smaller pieces and undergo further chemical and physical weathering. Decayed leaves, plants, and other organic material become part of the top layer of soil. These weathered, decayed remains of once-living things are known as humus. Most naturally formed topsoil is a mixture of small particles of weathered rock and humus.

Soils are constantly being formed in this way. They are also constantly being torn down by erosion.

When land contains a thick layer of good topsoil, it has a valuable resource. A centimeter (just under a half an inch) of good topsoil takes many years to be produced. It only takes a short time for erosion to destroy the topsoil of an exposed field. Fertilizers and chemicals can help improve soil, but they are expensive and still lack some of the benefits of good topsoil.

In many places the topsoil has been eroded away, but you probably found some evidence of topsoil as you examined the area in your investigation. The next layer of soil usually contains a mixture of gravel, sand, silt, and clay, although not all soils are alike. This loose soil layer may be shallow or it may extend hundreds of meters (a tenth of a mile). However, it will eventually meet a layer of solid rock such as hardened magma, cemented particles of rocks and sediment, or basement crust of the earth.

86

A view of Stratford, Texas, as a massive dust storm approaches (1935)

Dust Bowl Boundaries 1935-1940

Making Connections

The prairie soils in the north central part of the United States were, and still are, among the most fertile soils in the world. Many farmers were attracted to this region during the 1800s. It had a thick layer of humus, and the root system of the grasses kept the soil in place and held in moisture. By the late 1800s, new settlers moved farther west and south into a region known as the Great Plain. The soil there was also fertile, but there was less rain in the Great Plains than in the prairie soil region. Large numbers of farmers settled in the Plains and began to plow their fields and plant seeds. This worked well for many years, but in the 1930s there were several years of extremely low rainfall. The farmers continued to plow their fields and plant seeds, even though crops were poor. After a few years of continuing to plow their fields during a drought, the soil became very dry. The dust storms of the 1930s literally blew tons of loose dry soil, including most of the topsoil, to the east in huge black clouds. Much of the soil ended in the Atlantic Ocean. The land became known as the Dust Bowl.

Dig Deeper

- Find directions for making a compost bin. Many people who enjoy planting gardens and flowers find composts a good way to recycle dead leaves, grass clippings, and other organic materials. Composts are not practical at every location, and there may even be ordinances that regulate them. Try to interview a gardener who maintains a compost bin and find out more about their benefits and disadvantages.

- During the dust-bowl years of the 1930s in the United States, poor farming methods left the topsoil of many Great Plains states exposed. Combined with a period of drought, winds blew much of the topsoil away. Do some research on the dust bowls of the Great Plains. Try to find out what caused this and what happened to the farmers in the area. Try to learn about the soil conservation measures that were used to protect and restore the soil.

What Did You Learn?

1. Humus makes up a part of the top layer of soil. How is humus made?

2. What other substances are found in topsoil?

3. A layer of loose soil found below the topsoil usually contains a mixture of materials. What is generally found in this layer?

4. If someone dug deep through the topsoil and the loose soil below, what would they find next?

5. Suppose a farmer loses most of the topsoil in a field as a result of using poor conservation methods. Will the soil be replaced by natural processes if he doesn't plant another crop on it the next year?

6. Can adding a bag of fertilizer to a field (where topsoil has been lost) make it as good as it was before the topsoil was lost?

Pause and Think

One of the truly amazing facts about the earth is how water, landforms, soil, and other things necessary for life to continue are recycled. No matter how polluted a single body of water becomes, pure, fresh water is always returned to the earth every day through the water cycle.

Weathering is continuously breaking down mountains, rocks, and soils on the earth. These weathered particles get carried away to new places by water, wind, ice, and other kinds of erosion. But elsewhere, soil and rocks are being remade. Processes such as sedimentation, volcanic releases, and cementing are continuously building up mountains, rocks, and other landforms.

Earthquakes, volcanoes, storms, and other catastrophes produce daily changes in the earth. Yet the earth continues to make adjustments to these changes and remains a stable and orderly place for us to live.

The runoff from rain and melting ice, carrying little particles of sediment with it, goes into drainage ditches, streams, and rivers. Rivers carry sediment and deposit it anywhere the river slows or stops moving — such as the inside of a curve in a river, where the river empties into an ocean or lake, and where rivers are stopped by a dam.

Sediment is usually dumped by a river into an ocean in a fan-shaped area known as a delta. Under the right conditions, these sediments may eventually form sedimentary rock. This rock would then be exposed to weathering processes and be eroded again.

The amazing resiliency of the earth is something we can be very thankful for, but this does not excuse us from failing to take care of the air, water, soil, and other natural resources of the earth in reasonable and responsible ways.

87

WHAT DID YOU LEARN?

1. Humus makes up a part of the top layer of soil. How is humus made? *It is made of decayed, weathered tree leaves, limbs, or other once-living things.*

2. What other substances are found in topsoil? *Small pieces of weathered rocks are also found and mixed with humus.*

3. A layer of loose soil found below the topsoil usually contains a mixture of materials. What is generally found in this layer? *Gravel, sand, silt, and clay, although not all soils are alike*

4. If someone dug deep through the topsoil and the loose soil below, what would they find next? *Solid rock*

5. Suppose a farmer loses most of the topsoil in a field as a result of using poor conservation methods. Will the soil be replaced by natural processes if he doesn't plant another crop on it the next year? *No. The formation of soil is a process that takes many years.*

6. Can adding a bag of fertilizer to a field (where topsoil has been lost) make it as good as it was before the topsoil was lost? *No. It would help, but a bag of fertilizer can't provide the same benefits of topsoil.*

Investigate the Possibilities

Elementary Earth Science

THE EARTH
Its Structure & Its Changes

Student Journal

Tom DeRosa
Carolyn Reeves

Date:

The Activity:
Procedure and Observations

1. Make a diagram of your flattened orange peel and label the north and south poles, the equator line, the longitudinal lines, the latitudinal lines, and the continents.

Drawing Board:

Compare the orange peel "map" with a flat map of the world. Compare with a globe if one is available.

2. Are the areas close to the poles really as large as they are drawn on a flat map of the world? _____ _____

3. You only drew six longitudinal lines on your orange map. How many longitudinal lines are usually shown on a world map? _____ _____

The longitudinal lines on a globe get closer together as they get closer to the poles. The latitudinal lines on a globe make smaller and smaller circles as they get closer to the poles. Make careful observations of your flat map of the world.

4. Does the flat map show these changes in the longitudinal and latitudinal lines?_____ _____ _____

What Did You Learn ?

1. Is the International Date Line a longitudinal line or a latitudinal line?
   _____

2. What is the name of the starting longitudinal line that is designated as 0°?_____

3. Which lines go from the North Pole to the South Pole?
   _____

4. Which lines circle the earth and are parallel to the equator?
   _____

5. What part of the earth doesn't have four seasons?
   _____

6. Into how many times zones is the earth divided?
   _____

7. What is a GPS device? What can a GPS device in an automobile do?
   _____
   _____
   _____

Stumper's Corner ✎

1. _____
   _____
   _____

2. _____
   _____
   _____

Date:

The Activity:
Procedure and Observations

Part A. Follow the directions to make a model of the interior of the earth with modeling clay. Write the names of the parts that make up the earth on the slips of paper. Use your reference sources to find additional information about each layer. On the back of the slips of paper, write at least two facts that you have found about each layer. Glue each of the labels to a toothpick and put the toothpicks in the correct part in the clay model.

1. Put this information in your answer booklet.

Drawing Board:

2. Look for patterns in the temperature and pressure going from the crust to the inner core of the earth. Write what you find about changes in temperatures and pressures going from the crust to the inner core.

Part B. Follow directions for making a mixture of cornstarch and water.

3. What happens to the mixture as you stir in the additional cornstarch? _____ _____

4. What happens to the mixture when you stop stirring and let the mixture set? _____

Push your finger through the mixture slowly.

5. When you push your finger through the mixture slowly, does it have properties of a solid or a liquid? _____

Now pick up some of the mixture and make it into a ball. Observe what happens when you squeeze it and when you let it sit in your hand.

6. What happens when you squeeze the ball in your hands? _____ _____ _____

7. What happens when you stop squeezing it and let it sit in your hand? _____ _____

Dig Deeper

What Did You Learn ?

1. In which layer of the earth are solid rocks found that are not extremely hot? ____________

2. The materials in the lower part of the mantle are extremely hot, but they are thought to be in a plasma state in places. Under what conditions does this plasma behave like a solid? ____________

3. Under what conditions does material from the mantle rise up into the crust or even to the surface of the earth? ____________

4. Most scientists believe the core of the earth is made of what elements? ____________

5. The three main states of matter are solid, liquid, and gas. What is meant by a plasma state? ____________

6. Is the earth's crust thicker under the continents or under the oceans? ____________

7. Some geologists believe there are convection-type currents in parts of the earth's mantle and core. What are some ways these currents might affect the earth? ____________

8. What are some of the natural resources found in the earth's crust? ____________

Stumper's Corner

1. ____________

2. ____________

Date:

The Activity:
Procedure and Observations

Part A Now cut out the six landmasses from your map. Remember, Europe and Asia are shown as one landmass, called the Eurasian Plate.

Some geologists believe the continents were connected to each other at one time. To see if that is a reasonable idea, try to slide the pieces toward each other to see if they fit together. You can move them right and left as well as up and down.

1. How well were you able to make the continents fit together?

2. Compare your idea of what a single continent might have looked like with what some geologists think this land mass might have looked like.

Part B Use the map in the appendix to find the major plates that have been identified by geologists. (You may want to compare this map with another up-to-date reference.) Locate the Pacific Plate, North American Plate, South American Plate, Caribbean Plate, Nazca Plate, African Plate, Eurasian Plate, Indo-Australian Plate, and Antarctic Plate. You may also find some other plates.

3. Where do the Pacific Plate and the North American Plate meet? _____

4. What plate is the Nazca Plate pushing against? _____

5. What plates meet across the nation of Israel? _____

6. Name some of the islands that are located where the Eurasian Plate connects with a plate to the west. _____

7. The little country of Haiti (where a 7.0 earthquake struck in 2010) is located near the boundary of what two plates? _____

Try to find the direction in which the plates making up or touching the United States are moving.

8. What information did you find? _____

_____

Part C Lay a hardboiled egg on a towel on a flat surface. Use a marker to draw a circle, 4 centimeters (an inch and a half) in diameter, on the side of the egg that is facing up. Carefully push the knife through the eggshell, following the lines of the circle.

Place your thumb on the cutout eggshell and move the circle back and forth a few millimeters (just under a tenth of an inch). Allow the cutout shell to collide with the rest of the shell.

You are trying to make a model with the eggshells that represents the following:

Subduction - two plates collide and one plate slides under the other one	Seafloor spreading - one plate slides apart from the other one	Plates that slide and grind past each other	Plate collision and hills and mountains are pushed up between them

9. Make diagrams of each of the four situations above and briefly write what is happening in each. _____

_____

_____

1. In general, where do the Pacific Plate and the North American Plate meet? _____

2. What is happening where the Pacific Plate and the North American Plate meet along the coast of California? _____ _____ _____

3. Where is the Mid-Atlantic Ridge? _____ _____

4. Is the Mid-Atlantic Ridge a region of subduction or seafloor spreading? _____ _____

5. What happens frequently along or near the San Andreas Fault in western California? _____ _____

6. The Himalayan Mountains are thought to have formed when two crustal plates did what? _____ _____

7. What do geologists believe about a land mass known as Pangaea? _____ _____ _____ _____

Stumper's Corner

1. _____ _____ _____

2. _____ _____ _____

ACTIVITY 4

Investigation #4
Earthquake

Date:

The Activity:
Procedure and Observations

Part A Wear safety glasses or goggles as you do these investigations. Take a popsicle stick and hold it at both ends, using two hands. Slowly bend the stick, making an arch. Release the pressure on the stick and observe.

1. Describe what happens when the pressure on the stick is released. _____

Hold and bend the stick as you did before, but keep bending the ends toward each other until it breaks.

2. At what point did the stick break? _____
_____

Part B Follow directions for putting together the equipment for this investigation. Now begin to slowly and steadily pull the block across the sandpaper. The rubber band will begin to stretch, but continue to exert pressure with a slow steady pull until the block moves. (Caution: Observers should wear safety glasses and not get too close to the block, as it may move in an unpredictable way.)

3. Write your observations about how the block moves. Notice if it moves forward smoothly or if it surges forward suddenly. _____
_____

4. Measure the distance the block moves forward in centimeters. _____

5. Describe what happens to the "building" as the block moves. _____
_____

Observe the rubber band as you pull on the block.

6. What does the rubber band do during this investigation? _____
_____

Stretch the rubber band around two stationary objects. Pull back on one side and then quickly release it.

7. Does the rubber band continue to vibrate for a few seconds? _____
_____

Optional Repeat this investigation, but use fine sandpaper this time instead of coarse sandpaper.

8. Measure the distance the block moves forward in centimeters. _____

9. Compare what happened this time with what happened when using coarse sandpaper. _____
_____

Dig Deeper

Stumper's Corner

1. ______________________________

____

____

2. ______________________________

____

____

What Did You Learn ?

1. According to the elastic rebound theory, how do the rocks break loose from a position of tension — do they suddenly surge forward or do they gradually move forward? ______________________________

____

2. According to the elastic rebound theory, what often causes the ground to shake during an earthquake? ______________________________

____

3. An earthquake begins as locked-up sections of rocks break free. Is the tension on the rocks increased or decreased after an earthquake?

____

____

4. Sometimes buildings fall in during an earthquake. What are some of the things that play a big role in how well a building can withstand an earthquake? ______________________________

____

5. Where are earthquakes most prone to occur? ______________________________

____

6. What kind of crustal plate movement is occurring along the San Andreas Fault? ______________________________

7. Is the New Madrid fault system located at the boundary of two major crustal plates? ______________________________

8. One of the main earthquake belts in the earth is known as the "Ring of Fire." What seems to be happening all along the coastline of the Pacific Ocean to cause so many earthquakes? ______________________________

____

Date:

The Activity:
Procedure and Observations

Part A Refer to the chart in your textbook that gives information about some of the world's biggest earthquakes since 2001.

Get a copy of a world map you can mark on (from your teacher). Use the chart above to find information about recent major earthquakes. Use references to help you find places you don't know.

1. Mark the earthquakes by making a small red circle at or near each location on the map. Next to the red dot, write the date of the earthquake.

2. When you finish, look for patterns on the map about where most recent earthquakes have occurred. What patterns did you find? _____
   _____
   _____

Now look in a reference source to find the major tectonic plates on the earth. About eight major plates have been identified, along with a few other smaller plates.

3. Do you see a connection between the earthquakes on your map and places where plates meet? _____

4. About how many of the earthquakes on your map were located where two plates meet? _____

5. Color the regions with the most recent strong earthquakes yellow.

6. Do strong earthquakes always cause a large number of deaths? _____
   _____

Part B. Optional. Follow directions in your textbook for this investigation. Shake each house back and forth for several seconds. Move your hand 3 centimeters (a little over an inch) forward and 3 centimeters back as you shake the board. Shake the houses the same way each time.

7. Explain what happens to each of the two houses after they have been shaken. _____
   _____

8. Do you think the marshmallows (or marbles) provided a way to make the house less likely to fall in? _____
   _____

Dig Deeper

What Did You Learn ?

1. The strength of an earthquake is reported as a number from 0 to 10. What is this scale of numbers called? ______________________

2. What is the name of the instrument that is used to study and identify earthquakes? ______________________

3. About how often do earthquakes occur throughout the world — every day or about once a month? ______________________

4. Which earthquake is more powerful — one that measures 3 on the Richter Scale or one that measures 9 on the Richter Scale? __________

5. What is the advantage of designing a building that moves slightly on its foundation? ______________________
____

6. What are some of the major earthquakes that have struck the United States? ______________________
____

7. What are some ways the government of a country could reduce the deaths and damage caused by an earthquake? ______________________
____

8. Where are the two main earthquake belts in the earth? __________
____

Stumper's Corner

1. ______________________
____
____

2. ______________________
____
____

Pause and Think: ?

What are your thoughts about this? ______________________
____
____
____

Investigation #6
Volcanoes

Date:

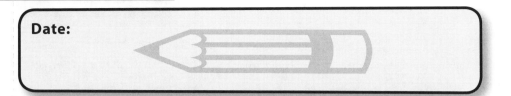

The Activity:
Procedure and Observations

Fill a few film canisters about 2/3 full of water, add ¼ of a piece of an effervescent tablet to each canister, and fit the cap on securely. Gently shake the canisters, wait several seconds, and observe what happens.

1. What happened to the canister caps? _____
   _____

2. Try to think of an explanation for what happened before you read "The Science Stuff."_____
   _____
   _____
   _____

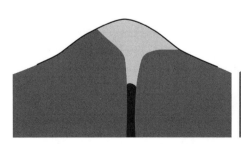

A "capped" volcano with cooled magma. See relation to instruction 1.

The rising magma bubbles up. Relate to instruction 2 concerning shaking the container.

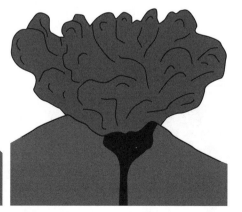

Pressure builds and blasts off cap. Relate this to instruction 3.

Dig Deeper

What Did You Learn ?

1. When magma reaches the surface of the earth, what is it called?

2. Were there warning signs that Mount St. Helens was about to erupt before May 18, 1980?

3. What has probably happened if a volcano suddenly erupts and forcefully throws out volcanic materials and rocks?

4. When did the island of Surtsey first appear?

5. Where are some of the most likely places to find volcanoes in the world?

6. When magma from the mantle is pushed up through cracks in the crust, does it always reach the surface of the earth?

7. Are volcanic eruptions always violent and explosive?

Stumper's Corner

1.

2.

Date:

The Activity:
Procedure and Observations

Part A. Flatten the clay into four rectangles, each about 7 x 15 centimeters (2.5 x 6 inches). Stack the layers of clay like you were making a sandwich and place them flat on a table. Carefully place a pencil under the middle of the stack and lift up a few centimeters. Place the pencil under one end of the stack and lift up a few centimeters. This represents what happens when pressure rises up from below.

1. Observe the shape and make a drawing of how the four layers look when held from the middle.

Drawing Board:

2. Make a drawing of how the four layers look when raised from one end.

Drawing Board:

Now take your stack of clay rectangles and lay them flat on the table. Slowly push in from both ends until the clay begins to buckle. This represents what happens when pressure is applied against the sides of plastic rock material.

3. Observe the shape and make a drawing of how the four layers look when pressure is applied against the sides of the rock material.

Drawing Board:

4. Did you observe any breaking of the layers of clay in either case? ___
_____
_____

Color the edges of the four layers of Styrofoam, making each piece a separate color. Stack the four layers of Styrofoam and lay them flat on the table. Repeat the activities above, but use Styrofoam instead of clay.

5. Write your observations about what happened when you tried to lift the stack of Styrofoam from the middle. _____
_____

6. Write your observations about what happened when you pushed on the layers until they buckled. _____
_____

7. What is the difference between the clay and the Styrofoam as pressure is applied to each? _____
_____
_____
_____

Dig Deeper

What Did You Learn

1. Give an example of a place in the United States where miles of flat, level layers of strata can be seen. ____________________

2. Give an example in the United States where extensive folding of sedimentary layers can be seen. ____________________

3. Name at least four ways in which sedimentary layers can be laid down in nature. ____________________

4. Which of the following processes is most likely to produce flat, level layers of sediment — glaciers, wind, water, or volcanic eruptions?

5. What are some things that can change flat, horizontal sedimentary layers in nature after they have been laid down? ____________________

6. What do we call breaks and cracks in large rock formations when rock on one side of the crack has slipped and moved? ____________________

Stumper's Corner

1. ____________________

2. ____________________

Pause and Think: Age of Rocky Mountains

What are your thoughts about this? ____________________

Date:

The Activity:
Procedure and Observations

(Caution: This activity should be adjusted if any students are diabetic or have peanut allergies. Adult supervision required.)

Follow directions in your textbook to make a model of extrusive and intrusive rocks with sedimentary layers of rocks. Now cut out a slice of the layers, being sure to cut through the center of one of the rolled wafers. Determine what each part of your model represents, and stick toothpicks in each part: red for hardened magma on the surface of the ground; green for hardened magma below the surface of the ground that formed between rock strata; yellow for hardened magma below the surface of the ground that formed through the strata in a (more or less) vertical position; blue for each original strata of rock.

Read The Science Stuff and identify extrusive rocks, intrusive sills, and intrusive dikes in the bread layers model.

1. What were the red toothpicks? _____

2. What were the green toothpicks? _____

3. What were the yellow toothpicks? _____

4. What were the blue toothpicks? _____

5. Which rocks were the oldest? _____

Draw two or more of your cut out slices. Label each part where you placed a toothpick.

6. Copy your labeled drawings in your student notebook.

Drawing Board:

What Did You Learn ?

1. How does magma travel from the mantle through the crust of the earth? _____
_____

2. When magma reaches the surface of the earth and hardens there, what kind of rock forms? _____

3. What are intrusive rocks?_____
_____

4. Are dikes and sills intrusive or extrusive rocks? _____
_____

5. Suppose you find an intrusive rock that has passed through several layers of stratified rocks. Which would be older — the intrusive rock or the stratified layers of rocks?_____
_____

6. How does the cooling rate affect the size of mineral crystals that form in intrusive magma? _____
_____
_____

✏ **Stumper's Corner**

1. _____
_____
_____

2. _____
_____
_____

Investigation #9
Mapping a Mountain

Date:

The Activity:
Procedure and Observations

Follow the directions in your textbook to mold the clay into the shape of a mountain that is 8 to 10 centimeters (3 to 4 inches) high. Make both a steep cliff and a gradual slope.

Continue to follow the directions to mark the contours of your mountain. There should be at least three levels marked off when you finish.

Carefully lift the "mountain" and remove the paper. Place the "mountain" back on the table and stand over it to view the three contour lines you made in the clay. Estimate the size and shape of each of the three contour lines and try to draw them on your paper within the outline of the bottom of the "mountain." Start with the lowest contour line and end with the highest. Each contour line you draw will be smaller than, and inside, the last one. This will be an estimate, so do the best you can.

Put the number "0" beside the largest contour line (the outline of the bottom of the "mountain." Measure the distance from the table to the first line in centimeters and write this number beside the next contour line. Measure the distance from the table to the third contour line and write this number. Repeat for each contour line.

1. Try to trace or recopy your four contour lines on a piece of paper.

2. Are all of your contour lines the same shape? _____

3. Are all of your contour lines the same distance apart? _____

4. Is the second contour line the same height above the table all the way around? _____

5. Compare your drawing to the topographic map below.

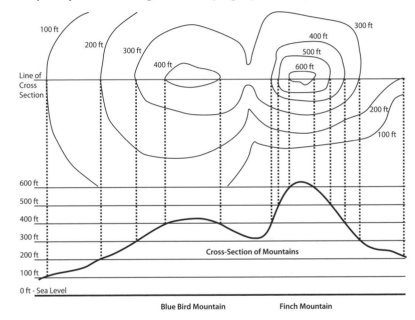

6. What is the highest elevation shown on the topographic map? _____
_____

7. Try to think of a reason why the contour lines are very close together on Finch Mountain and farther apart on Blue Bird Mountain. _____
_____
_____

8. Do any of the contour lines cross over each other? _____

9. How many hills are shown on the map? _____
_____

1. Complete the cross-section diagram of Hawk Mountain. (Half of it is already given.)

2. Complete the cross-section diagram of Martin Mountain.

3. How tall is Hawk Mountain? _____

4. How tall is Martin Mountain? _____

5. Each contour line represents how many feet of elevation? _____
   _____

6. Which mountain has steeper slopes (contour lines are closer together)? _____

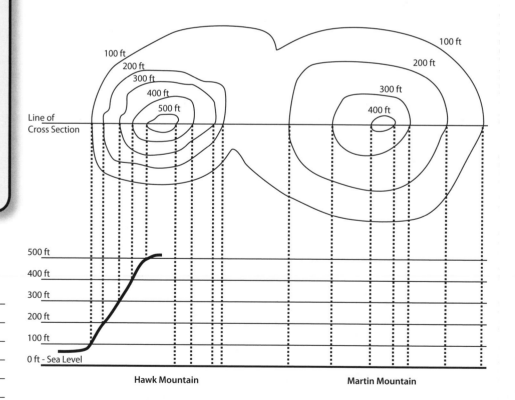

Stumper's Corner

1. _____
   _____
   _____

2. _____
   _____

Date:

The Activity:
Procedure and Observations

Follow the directions in your textbook to make three saturated solutions of alum and water. Tie a small washer to the end of each string and lower each washer into one of the small glasses. Tape the other end of the string to the outside of the glass or to the table.

Leave one glass undisturbed for a few hours (up to one day). Leave another glass undisturbed for a few days. Leave another glass undisturbed for a week or more.

When you are ready to examine the crystals, carefully remove the string and washer and place them on a dark piece of paper. Examine the crystals under a magnifying lens.

1. Draw one or more crystals each time you examine them.

Drawing Board:

2. What differences do you notice in the crystals from each glass? ____

____

____

Drawing Board:

Each time you examine the crystals, also look at a few small rocks with a magnifying lens.

3. What differences do you see in the rocks and the alum crystals? ____

____

____

Drawing Board:

1. Do all crystals have the same shape? _____

2. How can you tell the difference in a crystal and a rock? _____
   _____
   _____
   _____

3. When crystals form from conditions in which they grow rapidly, do they tend to be large or small? _____

4. Can the shape of crystals be used to help identify kinds of crystals?
   _____
   _____

5. What are crystalline structures that are found in nature called?
   _____
   _____

6. Do large alum crystals have the same shape as small alum crystals?
   _____
   _____
   _____

Stumper's Corner

1. _____
   _____
   _____

2. _____
   _____
   _____

Date:

The Activity:
Procedure and Observations

If you already have a rock and mineral collection, you're ahead of the game. If not, this is the time for you to start your own rock collection. Look around your house and neighborhood for loose rocks. Get several small plastic bags that you can close easily. Put each rock you find in a separate bag and label the bag with a permanent marker. Put your name, where you found it, the date, and the name of the rock if you know it. Make a note if you think the rocks were brought in from some other place and are not local rocks.

Some of the rocks you find will probably look like it contains the same material throughout. However, some rocks are made of pieces of minerals (crystals) from several different kinds of minerals that have been cemented together. Try to find some rocks that have little pieces of shiny things embedded in them. Examine these rocks under a magnifying lens.

1. Describe the shiny pieces in the rocks as much as you can. _____
   _____
   _____
   _____

Some of the more common minerals are shown in your textbook. Refer to these pictures if you don't have examples of these minerals.

If you have your own mineral collection, also refer to them.

2. Make careful observations of the pictures (or minerals) and list some ways minerals are alike. _____
   _____
   _____
   _____

3. Now list some ways minerals are different. _____
   _____
   _____
   _____

Look at the crystals in the pictures.

4. Compare the shapes of each. _____
   _____
   _____
   _____

Dig Deeper

What Did You Learn

1. When crystals are found in nature, what are they called? ____________

2. Why is granite classified as a rock instead of a mineral? ____________

3. Are minerals and crystals pure substances or mixtures of chemicals? ____________

4. Names the minerals found in granite. ____________

5. Name at least eight tests that can be done to identify minerals. ____________

6. What kinds of minerals make up about 95 percent of the minerals in the earth's crust? ____________

7. Name at least eight minerals that make up most of the rocks in the earth's crust. ____________

Stumper's Corner

1. ____________

2. ____________

Pause and Think: Day 3

What are your thoughts about this? ____________

Date:

The Activity:
Procedure and Observations

Part A Choose five rocks that seem to be different from each other. Include one limestone rock. Place your rocks on paper plates and number each rock. Now place each rock in a clear container of water, one at a time, and try to determine if the rocks change in any way. Leave each rock in the container for at least 20 seconds.

1. What happens? Record your observations for each rock in the chart.

Dry the rocks. Now place each rock, one at a time, in a clear container of vinegar. Try to determine if there are any changes in the rocks this time. Leave each rock in the container for at least 20 seconds.

2. What happens? Record your observations for each rock in the chart.

Dry the rocks, and do one more test. Hold a penny in one hand and firmly rub the edge of the penny across each rock. Notice if the penny makes a scratch mark on the rock that doesn't rub off with your fingers. Look at any scratch marks with a magnifying lens.

3. What happens? Record your observations for each rock in the chart.

Dry the rocks again and save them for the next lesson.

Rock #	Reaction to water	Reaction to vinegar	Can a penny scratch?
1			
2			
3			
4			
5			

Dig Deeper

What Did You Learn ?

1. What are three tests or observations that can help you identify limestone? ______________________

2. What is the name of the chemical in limestone? ______________

3. Would you be more likely to find fossils in limestone, marble, or granite rocks? ______________

4. Limestone often contains the fossilized remains of what kinds of animals? ______________

5. There are three main groups of rocks — sedimentary, metamorphic, and igneous. In which group would limestone be classified? ______________

   In which group would marble be classified? ______________

6. What conditions are generally necessary in order for metamorphic rocks to form? ______________

7. Limestone has many characteristics that make it a good building material. Why do builders often cover it with a sealant? ______________

Stumper's Corner

1. ______________________

2. ______________________

Rocks Have an ID

Date:

The Activity:
Procedure and Observations

Part A Choose 12 rocks from your collection that seem to be different from each other. Place a different rock in each of the holders of an empty egg carton. Number each rock by writing on the egg carton. The rocks should be dry. Firmly rub rock number one across a porcelain plate or the backside of the ceramic tile and observe the color of the dust it makes. The harder rocks will not leave a streak, but may make a scratch mark.

1. Look for any colored streak left by the rock. Record this in your data table.

2. Record the color of the rock and see if it is the same as the color of the streak.

3. Repeat with the rest of the rocks. Record the color of any streak left by each rock. Record "none" if a rock doesn't leave a streak.

Try to see if you can use any of your rocks to write on a piece of white paper. Try to see if you can use any of your rocks to write on a piece of dark-colored paper.

4. Write the numbers of any rocks that can write on white paper. Write the numbers of any rocks that can write on colored paper.

Try to scratch each rock with a copper penny.

5. Write the numbers of any rock that a penny can scratch.

Try to scratch each rock with an iron nail.

6. Write the numbers of any rock that an iron nail can scratch.

Use rock and mineral identification charts on the Internet or other references to help you identify or classify the rocks in your collection. This may be more difficult than you would expect, but tentatively write the names of what you think they are.

7. List tentative names of your rocks. Give the tests you did to try to identify them. Use the results of Part B to help you if you did this.

Test	1	2	3	4	5	6	7	8	9	10	11	12
Color of rock												
Streak test												
Write on white												
Write on colored												
Scratch by penny												
Scratch by nail												

Part B (Optional) Compare the hardness of different rocks.

Take the first two rocks in your collection and try to scratch one with the other. Keep the rock that is the hardest and set the other rock aside. (Put the rocks you set aside on a sheet of paper and keep up with their numbers.) Follow the directions in your textbook to complete this investigation. List the rocks you tested in order from the hardest to the least hard.

✎ **Dig Deeper**

✎ **Stumper's Corner**

1. _____
_____
_____

2. _____
_____
_____

What Did You Learn ❓

1. Explain how to do a streak test on a rock._____
_____
_____

2. Use the Mohs hardness scale to find one mineral that the mineral apatite can scratch and one mineral that it cannot scratch. _____
_____

3. Name the three big groups of rocks. _____
_____

4. Which group of rocks contains fossils?_____

5. Which kinds of rocks are formed when igneous rocks or sedimentary rocks are subjected to high temperatures and/or pressure?_____
_____

6. Which kind of rock forms from the cooling of magma underground before it reaches the surface of the earth?_____
_____

7. Name three or four sedimentary rocks. _____
_____

8. Basalt, granite, and obsidian are examples of what kind of rock?
_____

9. Marble is a metamorphic rock that is made from what sedimentary rock under conditions of great heat and pressure? _____
_____
_____

How Little Tiny Things Settle Out of Water to Become Rocks

Date:

The Activity: Procedure and Observations

Put sand, gravel, soil, and a few twigs in a quart jar until it is about halfway full. Then add water until the jar is about 2/3 full.

1. Predict which materials you think will settle to the bottom first and which ones will be on top after they have been stirred up._____
_____
_____

Place the lid on the jar and shake vigorously to mix. Place the jar on a flat surface and let it stand until all the materials are settled and the water becomes somewhat clear. (See directions in textbook.)

2. Observe and record the order in which the materials settled. Is this what you predicted would occur? _____
_____
_____

3. Make a drawing of what you observe. Be sure to label the layers that form.

Drawing Board:

Shake the jar again as you did before and place it on a flat surface. Wait 15 minutes.

4. Did you get about the same results that you did the first time? _____
_____
_____

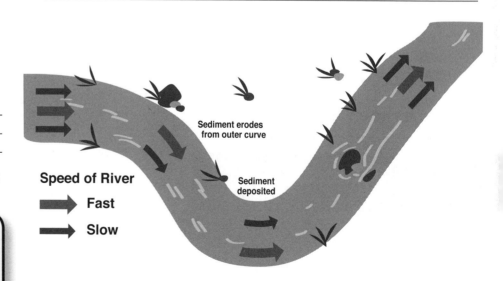

Sediment erodes from outer curve

Sediment deposited

Speed of River

➡ Fast

➡ Slow

Dig Deeper

Stumper's Corner

1. ______________________________

____

____

2. ______________________________

____

____

What Did You Learn ?

1. What term refers to materials, such as stones, sand, and silt, that are deposited by water? ______________________________

2. What is the process of depositing sediment out of water? ______________________________

3. What are two other things that can deposit sediments? ______________________________

4. Is the settling rate faster for larger/heavy materials or for smaller/lighter materials? ______________________________

5. Where are sand bars likely to be found — on the inside of a river curve where the water moves slowly or on the outside of a river curve where the water moves faster? ______________________________

6. The Mississippi River carries tons of sediment into the Gulf of Mexico every year. What is the area called where the river and the gulf meet? ______________________________

7. Why is most of the sediment deposited in this area rather than farther upstream? ______________________________

8. What is most likely to leave deposits of large boulders — water, wind, or glaciers? ______________________________

9. When sediment hardens or consolidates, what kind of rocks are formed? ______________________________

Date:

The Activity:
Procedure and Observations

Follow the directions in your textbook to make a model of how rain can cause erosion of soil. Hold the cup you punched holes in over the higher end of the pan and fill it with water. The falling water will act like raindrops hitting the dirt.

1. Observe the effects of the "rain" and record what you see. _____
_____
_____

Refill the cup with water each time it gets empty, and count the number of cups of water it took for all the dirt to be moved to the lower end of the pan. You can move the cup back and forth over the elevated end, but try to keep the distance above the dirt about the same.

2. How many cups of water were needed to move the dirt to the lower end of the pan? _____

Remove the wet dirt, dry the pans, and repeat the activity. This time add two more books to make the incline sleeper. Repeat the investigation.

3. How many cups of water were needed to move the dirt to the lower end of the pan this time?_____

4. Did you make any other observations? _____
_____
_____

Remove the wet dirt, dry the pans, and set everything up as in the last activity. This time place twigs and leaves or pine straw over the dirt. Repeat the investigation, except allow the same amount of water to fall as in the last activity.

5. What differences did you observe in the amount of erosion this time?
_____
_____

6. Did the twigs, leaves, or pine straw help to prevent erosion of the soil?_____
_____

How erosion works

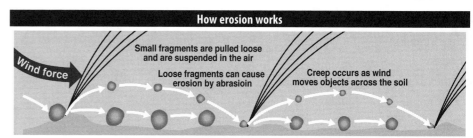

Wind Erosion — Wind can lift small fragments of rock or can cause these fragments to strike others, thus removing more particles.

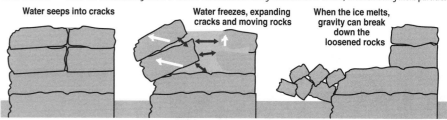

Ice and Gravity Erosion — When water freezes and then melts, it loosens stones, thus allowing gravity to eventually tear down the rock structures.

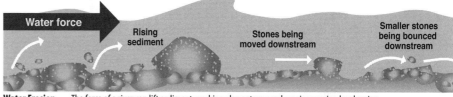

Water Erosion — The force of a river can lift sediment, pushing along stones and causing erosion by abrasion.

What Did You Learn ?

1. What is erosion? _____

_____

2. What things can cause erosion? _____

_____

3. How are erosion and sedimentation different? _____

_____

4. Under what conditions will erosion occur at a faster rate? _____

_____

5. How can erosion be slowed or stopped? _____

_____

6. Why is it important to take care of topsoil in an area? _____

_____

7. Use diagrams to show how a meandering river can change course after several years.

8. What is an oxbow lake? _____

_____

Stumper's Corner ✏️

1. _____

_____

_____

2. _____

_____

_____

Pause and Think: Hundreds of flood stories ?

What are your thoughts about this? _____

_____

_____

_____

Date:

The Activity:
Procedure and Observations

Part A. Physical Weathering Fill an aluminum soft drink can as full of cold water as you can. Place it in a container with a flat surface and put it in the freezer. Leave it for several hours until the water is frozen solid.

1. Describe what happened to the aluminum can. _____

_____

2. Try to give an explanation for what happened before you read "The Science Stuff." _____

_____

3. Compare your explanation with the explanation in the book.

Combine a handful of rocks in a disposable plastic jar with some hard candy. Screw the lid on the jar and shake the jar so that the rocks and candy alternately hit the lid and the bottom of the jar. Do this several times. Stop shaking when you see several broken pieces of candy or rocks. Examine the contents.

4. Look carefully at the broken fragments. Describe how they look. Draw a few of the fragments on a piece of paper.

5. Compare the broken fragments with the weathered rocks. Which fragments had the most sharp, angular shapes? _____

6. Were there more fragments of candy or more fragments of rocks?

_____

Put one or two drops of water on some of the candy fragments that have clumped together and let dry. This represents rock pieces that get cemented together.

7. Would cementing rock pieces together be an example of weathering or would it be the opposite of weathering? _____

Part B. Chemical Weathering Take some plain steel wool (which is made of iron) and dip it in a solution of salt water. Place the pad in an empty margarine tub or other container. Set it aside but continue to dip it in the salt water from time to time. After a few days, examine the steel wool again.

8. What color do you observe that was not present when you began this investigation? _____

9. Is the new substance hard or is it crumbly? _____

(Optional) You may want to repeat the investigation in which you put limestone rocks in a container with vinegar. This represents another example of chemical weathering by exposure to an acid.

10. Describe what happens. _____

_____

11. Describe the limestone rocks after you have removed them from the vinegar and washed them in water. _____

_____

Dig Deeper

Stumper's Corner

1. ______________________________
____
____

2. ______________________________
____
____

What Did You Learn ?

1. Suppose a rock breaks into several small pieces after a big rock falls on it. Is this an example of physical weathering or chemical weathering? Explain your answer.______________________________
____
____
____

2. What kind of shape would you expect newly broken rocks to have — sharp and jagged or smooth and rounded?______________________________
____

3. Give some examples of how rocks can undergo physical weathering.
____
____
____

4. Give some examples of how rocks can undergo chemical weathering.
____
____
____

5. Why do rocks sometimes break during freezing weather when water gets into cracks in the rocks?______________________________
____
____
____

6. What are some of the differences between iron and iron oxide? _____
____
____
____

Date:

The Activity:
Procedure and Observations

Find the halfway point on a couple of cans and make a mark there. Start with a ½ can of dry sand. Then add ½ can of rocks on top of the sand. Shake the can gently a few times.

1. Do you have a full cup of sand and rocks? _____

Now start with ½ can of rocks and add ½ can of dry sand on top of the rocks? Shake the can gently a few times.

2. Do you have a full cup of sand and rocks? _____

3. Try to think of a reason for the difference. _____
_____
_____

Pour the rocks and sand through a wire strainer to separate them to continue using them.

Fill a can with water and pour the water into a metric measuring cup.

4. Record the volume of the water in milliliters that an empty can will hold. _____
_____

Fill one can to the top with rocks. Fill another can to the top with water. Carefully pour as much of the water as possible over the rocks without letting the water spill over. Measure the water that is left in the can. Subtract from the volume of the full can to find how much water went into the spaces around the rocks.

5. Record the amount of water that went into the spaces around the rocks in milliliters. _____

Fill a can to the top with dry sand. Slowly add a can full of water to the sand without letting the water spill over. Measure the water that is left in the can. Subtract from the volume of the full can to find how much water went into the spaces around the particles of sand.

6. Record the amount of water that went into the spaces around the sand in milliliters. _____

Compare the amount of water that went into the spaces around the rocks with the amount that filled in around the sand.

7. Which held the most water — a can of rocks or a can of sand? _____
_____

Punch holes into a paper cup with a toothpick. Place the paper cup in a bowl that is 3 or 4 times larger than the paper cup. Place dry sand in the container around the paper cup. Slowly add ½ can of water to the sand. After a few minutes, observe what happens inside the paper cup.

7. Record your observations. _____
_____
_____
_____

1. When farmers dig wells in order to get water to drink, how deep do the wells have to be in order to reach water?_____
   _____

2. What might cause a water well to go dry? _____
   _____

3. The spaces between rocks, sand, or other earth materials are called what? _____

4. The amount of spaces between rocks, sand, or other earth materials compared to the total volume of the material is known as what?
   _____

5. What is a water aquifer?_____
   _____

6. Are oil and gas deposits found in porous or nonporous layers? _____
   _____

7. Are most oil and gas deposits found near the surface of the ground or deep below the ground?_____
   _____
   _____

8. When water has filled all the pores in a porous underground area, the area is said to be _____ with water.

9. Why is there a great need for people all over the world to find drinking water from clean wells rather than getting their drinking water from streams and shallow polluted wells? _____
   _____
   _____

Stumper's Corner

1. _____
   _____
   _____

2. _____
   _____
   _____

Investigation #18
Caves, Sinkholes, and Geysers

Date:

The Activity:
Procedure and Observations

Punch several holes in a disposable aluminum pie pan. Line the pan with a layer of clay like a piecrust, going up the sides as well as the bottom. Make sure there are no holes or cracks in the clay. Place pan and crust in a 350° oven for a few minutes or let the crust dry out overnight. Allow the pan to cool and then fill the clay piecrust with dirt or potting soil. Place the pie pan in a larger container to avoid spilling water. Add about a cup of water to the pie pan. Wait a minute or two and observe.

1. Does any of the water leak out of the pie pan?_____
   _____
   _____

Now take a plastic knife and make several cuts in the soil and layer of clay. Be sure to cut all the way through to the aluminum, but don't punch more holes in the pan. Wait another minute or two and observe.

2. Does any of the water leak out of the pie pan? _____
   _____
   _____

Porous soil

Non-porous layer

Porous pan (holes in the bottom)

Dig Deeper

Stumper's Corner

1. ______________________________

2. ______________________________

What Did You Learn ?

1. When there is a rain, part of the water becomes runoff. What happens to the rest of the water?______________________________

2. What determines how deep water soaks into the ground?______________________________

3. Give an example of how underground rocks can be weathered physically.______________________________

4. Give an example of how underground rocks can be weathered chemically. ______________________________

5. Briefly explain how stalactites and stalagmites form in caves. ______________________________

6. Is the water that erupts from a geyser in Yellowstone National Park hot or cold? ______________________________

7. Where does the water come from that is spewed out of a geyser? ______________________________

8. When is a sinkhole most likely to fall in — when there has been plenty of rain or during a drought?______________________________

Thinking About

Date:

The Activity:
Procedure and Observations

Pour about 3 centimeters (just over an inch) of water in a small plastic container. Place the container in a freezer until the water is frozen. Run warm water over the bottom of the container to loosen the block of ice. Sprinkle a little sand or dirt over the ice. Place some small rocks in another plastic container that is a little larger than the block of ice. Place the block of ice over the rocks and place a few more rocks on top of the ice. Wait about three minutes and then return the ice and rocks to the freezer. Wait at least 30 minutes before examining the rocks and ice.

1. Predict what you think will happen to the rocks. _____
    _____
    _____

Remove the ice and rocks from the freezer and place them on a paper towel. Make careful observations of what you see. _____
_____
_____
_____

Drawing Board:

2. Did the rocks underneath the ice stick to the ice? _____

3. Did the rocks seem to be attached to the ice? _____

4. What did you observe about the rocks on top of the ice? _____
    _____
    _____

5. Make a diagram showing how the rocks were attached to the ice.

Drawing Board:

6. Was this what you predicted would happen? _____

Now put the rocks and ice in one end of a large, flat container. Tilt the container a few degrees and let the ice melt.

7. Describe what you observe after the ice has melted. _____
    _____
    _____

8. Make a hypothesis about what the rocks would do to the land if this were a large heavy glacier that was moving. _____
    _____
    _____
    _____

What Did You Learn ?

1. How do glaciers "pluck" rocks as they move? _____
   _____
   _____

2. Are glaciers able to transport rocks and other materials from place to place? _____
   _____

3. Are glacial deposits primarily pushed ahead of moving glaciers or moved by the process of plucking? _____
   _____

4. Are till deposits and moraines laid down by glaciers, wind, or water?
   _____

5. Briefly describe the two kinds of glaciers._____
   _____
   _____

6. Explain how glaciers cause erosion (or abrasion). _____
   _____
   _____

7. Why do rocks attached to glaciers start out jagged and end up being worn down and rounded?_____
   _____
   _____

Stumper's Corner

1. _____
   _____
   _____

2. _____
   _____
   _____